Dominik Schultes

Route Planning in Road Networks

Dominik Schultes

Route Planning in Road Networks

Accurate, Flexible, Efficient

VDM Verlag Dr. Müller

Imprint

Bibliographic information by the German National Library: The German National Library lists this publication at the German National Bibliography; detailed bibliographic information is available on the Internet at http://dnb.d-nb.de.

Cover image: www.purestockx.com

Publisher:
VDM Verlag Dr. Müller Aktiengesellschaft & Co. KG, Dudweiler Landstr. 125 a, 66123 Saarbrücken, Germany,
Phone +49 681 9100-698, Fax +49 681 9100-988,
Email: info@vdm-verlag.de

Produced in USA and UK by:
Lightning Source Inc., La Vergne, Tennessee, USA
Lightning Source UK Ltd., Milton Keynes, UK
BookSurge LLC, 5341 Dorchester Road, Suite 16, North Charleston, SC 29418, USA

ISBN: 978-3-8364-8398-8

Für meine Eltern.

Abstract

Computing optimal routes in road networks is one of the showpieces of real-world applications of algorithmics. In principle, we could use Dijkstra's algorithm—the 'classic' solution from graph theory. But for large road networks this would be far too slow. Therefore, there is considerable interest in speedup techniques, which typically invest some time into a preprocessing step in order to generate auxiliary data that can be used to accelerate all subsequent route planning queries.

Following the paradigm of *algorithm engineering*, we design, implement, and evaluate three highly-efficient and provably accurate point-to-point route planning algorithms—all of which with different benefits—and one generic many-to-many approach, which computes for given node sets S and T the optimal distances between all node pairs $(s,t) \in S \times T$ in a very efficient way. The evaluation is done in an extensive experimental study using large real-world road networks with up to 33 726 989 junctions.

Highway hierarchies exploit the inherent hierarchical structure of road networks and classify roads by importance. A point-to-point query is then performed in a bidirectional fashion—forwards from the source and backwards from the target—, disregarding more and more less important streets with increasing distance from source or target. *Highway-node routing* is a related bidirectional and hierarchical approach. Its conceptual simplicity and fast preprocessing allows the implementation of update routines that are able to react efficiently to unexpected events like traffic jams. *Transit-node routing* provides extremely fast query times by reducing most requests to a few table lookups, exploiting the observation that when driving to somewhere 'far away', the current location is always left via one of only a few 'important' traffic junctions. Our generic *many-to-many* algorithm can be instantiated based on certain bidirectional route planning techniques, for example, highway hierarchies or highway-node routing. It computes a complete $|S| \times |T|$ distance table, basically performing only $|S|$ forward plus $|T|$ backward queries instead of $|S|$ times $|T|$ bidirectional queries.

Among all route planning methods that achieve considerable speedups, we currently provide the one with the fastest query times, the one with the fastest preprocessing, and the one with the lowest memory requirements.

Acknowledgements

I would like to thank my PhD advisor Peter Sanders for the numerous inspir-ing discussions, his encouragement and support. I enjoyed working on joint papers with Holger Bast, Daniel Delling, Stefan Funke, Sebastian Knopp, Domagoj Matijevic, Peter Sanders, Frank Schulz, and Dorothea Wagner. The persistent support by my current and former colleagues was very help-ful, in particular by Norbert Berger, Roman Dementiev and Johannes Sin-gler w.r.t. technical issues and by Anja Blancani and Sonja Seitz w.r.t. organ-isational issues. Timo Bingmann's and Jiwei Wang's work on visualisation tools was of great value for the development process and the preparation of presentations. In addition to the already named people, I would like to thank Reinhard Bauer, Andrew Goldberg, Joaquim Gromicho, Martin Holzer, Ste-fan Hug, Riko Jacob, Ramon Lentink, Leo Liberti, Dietfrid Mali, Giacomo Nannicini, Paul Perdon, Dennis Schieferdecker, and Renato Werneck for interesting discussions. Veit Batz, Daniel Delling, and Martin Holzer proof-read parts of my thesis. The companies PTV AG in Karlsruhe and ORTEC in Gouda, the Netherlands, provided real-world road networks that had an important impact on the quality of the experimental evaluation of the con-sidered route planning techniques. Last but not least, I would like to thank Rolf Möhring for his willingness to examine my thesis. My work has been largely supported by DFG grant SA 933/1-3. The photo on the back cover has been taken by Fotostudio Becker, Karlsruhe.

Contents

1

Introduction

1.1 Motivation

Computing best possible routes in road networks from a given source to a given target location is an everyday problem. Many people frequently deal with this question when planning trips with their cars. There are also many applications like logistic planning or traffic simulation that need to solve a huge number of such route queries.

Current commercial solutions usually are slow or inaccurate. The gathering of map data is already well advanced and the available road networks get very big, covering many millions of road junctions. Thus, on the one hand, using simple–minded approaches yields very slow query times. This can be either inconvenient for the client if he has to wait for the response or expensive for the service provider if he has to make a lot of computing power available. On the other hand, using aggressive heuristics yields inaccurate results. For the client, this can mean a waste of time and money. For the service provider, the developing process becomes a difficult balancing act between speed and suboptimality of the computed routes. Due to these reasons, there is a considerable interest in the development of more *efficient* and *accurate* route planning techniques.

1.1.1 The Shortest-Path Problem

A road network can easily be represented as a *graph*, i.e., as a collection of nodes V (junctions) and edges E (road segments) where each edge connects two nodes. Each edge is assigned a weight, e.g. the length of the road or an estimation of the time needed to travel along the road. In graph theory, the computation of *shortest*[1] *paths* between two nodes is a classical problem. Actually, we can distinguish between several variants of this problem:

- *point-to-point*: compute the shortest-path length from a given source node $s \in V$ to a given target node $t \in V$;

- *single-source*: for a given source node $s \in V$, compute the shortest-path lengths to all nodes $v \in V$;

- *many-to-many*: for given node sets $S, T \subseteq V$, compute the shortest-path length for each node pair $(s, t) \in S \times T$;

- *all-pairs*: a special case of the many-to-many variant with $S := T := V$.

In this thesis, we concentrate on the point-to-point (Chapters 3, 4, and 6) and the many-to-many variant (Chapter 5). Optionally, we might want to compute not only the shortest-path length, but also a description of the shortest path itself.

From a worst-case perspective, the problem has largely been solved in 1959 by Dijkstra [20], who gave an algorithm that solves the single-source shortest-path problem using $O(m+n)$ priority queue operations for a graph $G = (V, E)$ with n nodes and m edges.

1.1.2 Speedup Techniques

In practice, when we deal with very large road networks and the point-to-point problem, the running times of Dijkstra's algorithm are not satisfying. There are several aspects that suggest that we can do better:

1. In a sense, Dijkstra's algorithm is an overkill since it computes the shortest paths from a given node s to *all* nodes $v \in V$ and not only to *one* given node t. For the point-to-point problem, this can be improved by stopping Dijkstra's algorithm as soon as the shortest path to t is found, but still the shortest paths from s to all nodes v that are closer to s than t are determined (Figure 1.1).

---

[1]Note that, depending on the chosen edge weights, 'shortest' can refer not only to 'spatial distance', but also, for instance, to 'travel time'.

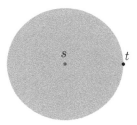

Figure 1.1: Schematic representation of the search space of Dijkstra's algo–rithm.

2. In many applications, we have to compute a lot of point–to–point queries on the *same* road network. Therefore, it can pay to invest some time for a *preprocessing* step that generates auxiliary data that can be used to accelerate all subsequent queries.

3. We do not deal with general graphs, but with road networks, which have certain properties. For instance, it is quite unusual for a node in a road network to have degree five or more, i.e., a road network is a very *sparse* graph. Furthermore, road networks are almost *planar* (because there are only a few bridges and tunnels in comparison to the total number of road segments) and usually a *layout* is given, i.e., the geographic coordinates of each node are known. Moreover, road networks exhibit *hierarchical properties*: for example, there are 'more important' streets (e.g. motor–ways) and 'less important' ones (e.g. urban streets).

These observations can be exploited to design *speedup techniques* that achieve considerably better query times than Dijkstra's algorithm when ap–plied to real–world road networks. There are several requirements that such a speedup technique should ideally fulfil:

- The query times should be as fast as possible.
- The result should be *accurate*, i.e., a provably optimal[2] path should be computed.
- The method should be *scale-invariant*, i.e., it should be optimised not only for long paths. In other words, the running time of the computation

---

[2]w.r.t. the available data

of a shortest path (e.g. from Karlsruhe to Saarbrücken) in a large graph (e.g. Western Europe) should be not much higher than the running time of the same computation in a smaller graph (e.g. Germany).

- If the approach uses some preprocessing, it should be sufficiently fast so that we can deal with very large road networks.

- Precomputed auxiliary data should occupy only a moderate amount of space.

- Updating some edge weights (e.g., due to a traffic jam) or replacing the entire cost function (e.g., switching to a different speed profile yielding different travel time estimates) should be supported.

Often these requirements are at conflict with each other. For example, faster query times might require larger preprocessing times and a larger memory consumption. The challenge is to find a method that represents a good compromise between all these requirements.

1.2 Related Work

1.2.1 Classical Results and Simple Techniques

Dijkstra's Algorithm [20] maintains an array of *tentative distances* for each node. The algorithm *visits* (or *settles*) the nodes of the road network in the order of their distance to the source node and maintains the invariant that the tentative distance is equal to the correct distance for visited nodes. When a node u is visited, its outgoing edges (u, v) are *relaxed*: the tentative distance of v is set to the length of the path from s via u to v provided that this leads to an improvement. Dijkstra's algorithm can be stopped when the target node is visited. The size of the search space is $O(n)$ and $n/2$ nodes on the average. We will assess the quality of route planning algorithms by looking at their *speedup* compared to Dijkstra's algorithm, i.e., how many times faster they can compute shortest-path distances.

Priority Queues. The main focus of *theoretical* work on shortest paths has been how to reduce or avoid the overhead of priority queue operations. The original version of Dijkstra's algorithm [20] runs in $O(n^2)$. This bound has been improved several times, e.g., to $O(m \log n)$ using binary heaps [99], $O(m + n \log n)$ using Fibonacci heaps [24], $O(m \log \log n)$ [84, 87], and $O(m+n \log \log n)$ using a sophisticated integer priority queue [89, 91] that supports *deleteMin* operations in $O(\log \log n)$ and all other operations in constant time. For integer edge weights in a range from 0 to C, Dial proposed an $O(m + nC)$ algorithm using buckets [19]. This bound has been improved to $O(m \log \log C)$ [93], $O(m + n\sqrt{\log C})$ [2], and $O(m + n \log \log C)$ [89, 91]. Linear time algorithms for the single–source shortest–path problem have been presented for *planar* [48] and *undirected* graphs [85, 86]. Meyer [59] gives an algorithm that works in linear time with high probability on an arbitrary directed graph with random edge weights uniformly distributed in the interval $[0, 1]$. Similar results have been obtained by Goldberg [27], whose algorithm is superior w.r.t. the worst–case bound for integer edge weights.

Experimental studies [13] indicate that in *practice* even very simple priority queues like binary heaps only induce a factor 2–3 overhead compared to highly tuned ones. In particular, it does not pay to accelerate *decreaseKey* operations since they occur comparatively rarely in the case of sparse road networks. In addition, our experiments indicate that the impact of priority queue implementations diminishes with advanced speedup techniques since these techniques at the same time introduce additional overheads and dramatically reduce the queue sizes.

Bidirectional Search executes Dijkstra's algorithm simultaneously for–wards from the source s and backwards from the target t (Figure 1.2). Once some node has been visited from both directions, the shortest path can be de–rived from the information already gathered [15]. In a road network, where search spaces will take a roughly circular shape, we can expect a speedup of around two—one disk with radius $d(s, t)$ has twice the area of two disks

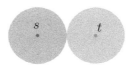

Figure 1.2: Schematic representation of the search space of the bidirectional version of Dijkstra's algorithm.

with half the radius. Many more advanced speedup techniques use bidirectional search as an optional or sometimes even mandatory ingredient.

Complete Distance Table. An extreme case would be to precompute all shortest paths. This allows constant time queries, but is prohibitive for large graphs due to space and time constraints. Still, it turns out that for some hierarchical approaches, this simple technique can be very useful when applied to the highest level of a hierarchy of networks.

1.2.2   Goal-Directed Search

Geometric A^* Search. A^* Search [35], a technique from the field of Artificial Intelligence, is a *goal-directed* approach, i.e., it adds a sense of direction to the search process. For each vertex, a lower bound on the distance to the target is required. In each step of the search process, the node v is selected that minimises the tentative distance from the source s plus the lower bound on the distance to the target t. This approach can be combined with bidirectional search [66]. The performance of the A^* search depends on a good choice of the lower bounds. If the geographic coordinates of the nodes are given and we are interested in the *shortest* (and not in the fastest) path, the Euclidean distance from v to t can be used as lower bound. This leads to a simple, fast, and space-efficient method, which, however, gives only small speedups. It gets even worse if we want to compute fastest paths. Then, we have to use the Euclidean distance divided by the fastest speed possible on *any* road of the network as lower bound. Obviously, this is a very conservative estimation. Goldberg et al. [29] even report a *slow-down* of more than a factor of two in this case since the search space is not significantly reduced but a considerable overhead is added.

Heuristic A^* Search. In the last decades, commercial navigation systems were developed which had to handle ever more detailed descriptions of road networks on rather low-powered processors. Vendors resolved to heuristics still used today that do not give any performance guarantees. One heuristic is A^* search with *estimates* on the distance to the target rather than lower bounds.

Landmark-Based A^* Search.   In [28, 29, 32], the ALT algorithm is presented that is based on $\underline{A}^*$ search, $\underline{L}andmarks$, and the $\underline{T}$riangle inequality. After selecting a small number of landmarks, for all nodes v, the distances $d(v, \mathcal{L})$ and $d(\mathcal{L}, v)$ to and from each landmark $\mathcal{L}$ are precomputed. For nodes v and t, the triangle inequality yields for each landmark $\mathcal{L}$ two lower bounds $d(\mathcal{L}, t) - d(\mathcal{L}, v) \leq d(v, t)$ and $d(v, \mathcal{L}) - d(t, \mathcal{L}) \leq d(v, t)$. The maximum of these lower bounds is used during an A^* search. For random queries, using 16 landmarks suffices to achieve a speedup factor of around 27 in the Western European road network consisting of about 18 million nodes. However, the landmark method needs a lot of space—two distance values for each node-landmark pair. It is also likely that for real applications each node will need to store distances to different sets of landmarks for global and local queries. Hence, landmarks have fast preprocessing and reasonable speedups, but consume too much space for very large networks. In Section 1.2.4, we will see that there is a way to reduce the memory consumption by storing landmark distances only for a subset of the nodes.

In [28] it is briefly mentioned that in case of an edge weight *increase*, the query algorithm stays correct even if the landmarks and the landmark distances are not updated. To cope with drastic changes or edge weight decreases, an update of the landmark distances is suggested. In [18], these ideas are pursued leading to an extensive experimental study of landmark–based routing in various dynamic scenarios.

Precomputed Cluster Distances.   The $\underline{P}$recomputed $\underline{C}$luster $\underline{D}$istances (PCD) technique [57] also uses precomputed distances for goal-directed search, yielding speedups comparable to ALT, but using less space. The network is partitioned into clusters and the shortest connection between any pair of clusters is precomputed. Then, during a query, upper and lower bounds can be derived that can be used to prune the search.

Signposts.   Another goal-directed technique is to precompute for each edge 'signposts' that support the decision whether the target can possibly be reached on a shortest path via this edge. During a query, only promising edges have to be considered.

Geometric Containers. A concrete instantiation of this general idea is the *geometric containers* approach [78, 79, 94, 98]. For each edge e, the set $S(e)$ is determined that contains all nodes that can be reached on a shortest path starting with e. Then, a simple geometric container $C(e)$ (e.g., a rectangular bounding box) is computed that contains at least all elements of $S(e)$. During the execution of Dijkstra's algorithm, an edge e can be ignored if the target node lies outside $C(e)$. While this approach exhibits a good query performance, the preprocessing step requires a very expensive all–pairs shortest–path computation so that no experimental results for the largest publicly available road networks have been published.

In [96], it is discussed how to modify geometric containers in order to react to edge weight changes.

Edge Flags (also called *arc flags*) [54, 52, 60, 61, 53, 55, 36] repre–sent a different instantiation of the general 'signpost idea'.[3] The graph is partitioned into k regions. For each edge e and each region r, one flag is computed that indicates whether e lies on a shortest path to some node in region r. Dijkstra's algorithm can take advantage of the edge flags: edges have to be relaxed only if the flag of the region that the target node belongs to is set. Obviously, inside the target region, the 'signposts' provided by the edge flags get less useful. This problem can be avoided by performing a bidirectional query so that forward and backward search can meet some–where in the middle.

While the query algorithm is very simple, the preprocessing process is more challenging. A naive procedure would have to perform an all–pairs shortest–path computation in the complete graph. A considerably better method reduces this effort to performing Dijkstra searches only from nodes that are adjacent to some node in a different region. A further improvement gets by with only one (though comparatively expensive) search for each re–gion. Using this most advanced preprocessing technique, Hilger [36] is able to preprocess the Western European road network with about 18 million nodes in about 17 hours achieving query times that are several thousands times faster than Dijkstra's algorithm. Note that an all–pairs computation,

---

[3]Note that geometric containers and edge flags have been developed independently of each other. However, since they share a common idea, we decided to subordinate both methods to a more general 'signpost approach'.

which would be required for the naive procedure (or for precomputing geo–
metric containers), would take more than six years on the same network.

The space consumption of the edge flag method can be reduced by ex–
tending it to a multi–level approach [60, 61] or by exploiting the fact that
many edges are associated with the same flags (already briefly mentioned in
[54] and extensively studied in [36]).

Particular advantages of the edge flag approach are the simplicity of
the query algorithm, the good query performance, the low memory require–
ments, and the fact that a complete description of the shortest path can be
derived for free (while some other approaches including ours have to un–
pack a contracted representation of the shortest path). Disadvantages are
the still comparatively slow preprocessing times, which make attempts to
deal with changing edge weights difficult, and the somehow limited query
performance in case of medium–range queries (e.g., between nodes that are
not very close, but still in the same region or in adjacent regions).

A comparison with our approaches can be found in Section 7.10.1. Very
recent developments cover combinations of edge flags with hierarchical
approaches—see Section 1.2.4. Moreover, using ideas from the edge flag
method could be used to further speed up our transit–node routing approach
(Section 6.5).

1.2.3 Hierarchical Approaches

Separators.   Road networks are almost planar: compared to the total num–
ber of road segments, the number of bridges and tunnels is very small.[4]
Therefore, techniques developed for planar graphs will often also work for
road networks. Such techniques often partition the given graph, exploiting
the Planar Separator Theorem [56], which states that for any planar graph
with n nodes, there is a node set of size $O(\sqrt{n})$—a so–called separator—
whose deletion leaves two components consisting of at most $2n/3$ nodes
each.

Various approaches use the following basic idea: They recursively parti–
tion the graph into several pieces. For each piece, they compute the shortest

---

[4]One widely used test instance, the US road network based on the TIGER/Line Files [92]
(cp. Section 7.2.2), is even planarised, i.e., the data set wrongly contains a junction at points
where a bridge/tunnel crosses over/under a street.

paths between all border nodes. Then, a shortest path search that passes through a certain piece need not consider all nodes within the piece, but can directly jump from border to border. For example, this idea can be recognised in [25], where Fuchs et al. distinguish between a fine and a coarse network representation. Another example is the construction and usage of the so-called *dense distance graph*: using $O(n \log^2 n)$ preprocessing time, query time $O(\sqrt{n} \log^2 n)$ can be achieved [21, 22, 49] for directed planar graphs with nonnegative edge weights; in a dynamic scenario, queries can be performed and edge weights can be updated in $O(n^{2/3} \log^{5/3} n)$ time per operation. A third example is [23], where Flinsenberg relies on the same basic idea and introduces modified versions of the A^* algorithm in order to compute routes for time-independent, time-dependent, and stochastic time-dependent scenarios.

The Separator-Based Multi-Level Method [78, 79, 80, 77, 37, 38] is a fourth example for an approach that uses the above mentioned basic idea. Out of several existing variants, we mainly refer to [37, basic variant]. For a graph $G = (V, E)$ and a node set $V' \subseteq V$, a *shortest-path overlay graph* $G' = (V', E')$ has the property that E' is a minimal set of edges such that each shortest-path distance $d(u, v)$ in G' is equal to the shortest-path distance from u to v in G. In the separator-based approach, V' is chosen in such a way that the subgraph induced by $V \setminus V'$ consists of small components of similar size. The overlay graph can be constructed by performing a search in G from each separator node that stops when all neighbouring separator nodes have been found. In a bidirectional query algorithm[5], the components that contain the source and target nodes are searched considering *all* edges. From the border of these components, i.e., from the separator nodes, however, the search is continued considering only edges of the overlay graph. By recursing on G', this idea is generalised to multiple levels. Speedups around ten are reported for railway transportation problems [80] and for road networks [98] that contain mostly nodes with degree two. In a more recent paper [38], speedups up to a factor of 52 are obtained for a

---

[5]In [37], the query algorithm is presented in two stages: first, determine the subgraph that should be searched; second, perform the search. We prefer to give a description with only one stage, which is simpler and virtually equivalent to a fully bidirectional variant of the original algorithm. Furthermore, analogies to highway-node routing (Chapter 4) will become more visible.

medium–sized road network. A limitation of this approach is that the graphs at higher levels become much more dense than the input graphs, thus limit–ing the benefits gained from the hierarchy. Also, computing small separators can become quite costly for large graphs. Closely related approaches have been suggested in [44, 45, 46, 47].

Bauer [6] observes that if the weight of an edge within some compo–nent C changes, we do not have to repeat the complete construction process of G'. It is sufficient to rerun the construction step only from some separator nodes at the boundary of C. No experimental evaluation is given.

In a theoretical study on the dynamisation of shortest–path overlay graphs [12], an algorithm is presented that requires $O(|V'|(n + m) \log n)$ preprocessing time and $O(|V'|(n+m))$ space, which seems impractical for large graphs.

One major part of this thesis is *highway-node routing* (Chapter 4), a route planning technique that is related to the separator–based multi–level method.

Thorup's Oracle [88, 90] is a different separator–based and hierar–chical approach. In a planar graph with integer edge weights in a range from 0 to C, queries accurate within a factor $(1 + \varepsilon)$ can be answered in time $O(\log \log(nC) + 1/\varepsilon)$ using $O(n(\log n)(\log(nC))/\varepsilon)$ space and $O(n(\log n)^3 (\log(nC))/\varepsilon^2)$ preprocessing time. Recently, this approach has been efficiently implemented and experimentally evaluated on a road net–work with one million nodes [64]. While the query times are very good (less than $20\,\mu s$ for $\epsilon = 0.01$), the preprocessing time and space consump–tion are quite high (2.5 hours and 2 GB, respectively).

Reach-Based Routing. Let $R(v) := \max_{s,t \in V} R_{st}(v)$ denote the *reach* of node v, where $R_{st}(v) := \min(d(s, v), d(v, t))$. Gutman [34] observed that a shortest–path search can be pruned at nodes with a reach too small to get to source or target from there. Speedups up to ten are reported for graphs with about 400 000 nodes using more than two hours preprocessing time. The basic approach was considerably strengthened by Goldberg et al. [26, 30, 31], in particular by a clever integration of *shortcuts* [69, 70], i.e., single edges that represent whole paths in the original graph.

A dynamic version that handles a set of edge weight changes is pre–sented in [6]. The basic idea is to rerun the construction step only from nodes within a certain area, which has to be identified first. So far, the concept of shortcuts, which is important to get competitive construction and query times, has not been integrated in the dynamic version. No experimental evaluation for the dynamic scenario is given in [6].

Heuristic Approaches. As an alternative to heuristic A^* search (Section 1.2.2), commercial navigation systems often use heuristic hierarchical approaches [41, 43], which perform bidirectional searches: While the forward/backward search is inside some *local area* around source/target, all roads of the network are considered. Outside these areas, however, the search is restricted to some *highway network* consisting of the 'important' roads. This general idea can be iterated and applied to a hierarchy consisting of several levels. The crucial point is the definition of the highway network. The heuristic approaches use a definition that is based on a classification of the streets according to their type (motorway, national road, regional road, ...). Such a classification requires manual tuning of the data and a delicate trade-off between speed and suboptimality of the computed routes.

Highway Hierarchies. Inspired by the just mentioned heuristic approaches, we developed exact *highway hierarchies* [75, 69]. Instead of blindly relying on the road types, we classify nodes and edges fully automatically in a preprocessing step in such a way that all shortest paths are preserved. By this means, we win not only exactness, but also greater speed since we can build high-performance hierarchies consisting of many levels without worrying about the quality of the results.

The local area is defined to consist of the H closest nodes, where H is a tuning parameter. Then, an edge $(u, v) \in E$ has to belong to the highway network if there are nodes s and t such that (u, v) is on some shortest path from s to t, v is not within the H closest nodes from s, and u is not within the H closest nodes from t. The resulting highway network can be pruned by removing isolated nodes and trees attached to a biconnected component, and by replacing paths consisting only of nodes with degree two (so-called 'lines') by single shortcut edges. After that, the construction process can be iterated. A schematic representation of the search space is given in Figure 1.3.

Figure 1.3: Schematic representation of the search space of the highway hierarchies approach.

Highway hierarchies are the first speedup technique that was able to handle the largest available road networks giving query times measured in milliseconds. There are two main reasons for this success: First, the road network shrinks in a geometric fashion from level to level and remains sparse, i.e., levels of the highway hierarchy are in some sense *self similar*. Second, preprocessing can be done very efficiently, using limited local searches starting from each node. Preprocessing is also the most nontrivial aspect of highway hierarchies. In particular, long edges (e.g. long-distance ferry connections) make simple-minded approaches far too slow. Instead, we use fast heuristics that compute a supergraph of the highway network.

In this thesis (Chapter 3), we present a greatly improved version of highway hierarchies.

Jacob and Sachdeva [42] experimented with different node numberings in order to obtain an I/O-efficient layout. They achieved a speed-up factor of around 1.3 compared to the default layout. In an experimental study [8], Bauer et al. apply highway hierarchies (and many other speedup techniques) to various types of graphs and not only to road networks. Their results indicate that highway hierarchies work (reasonably) well on condensed and time-expanded long-distance railway networks, unit disk graphs, and 2-dimensional grid graphs, while they fail on some local-traffic railway networks, higher dimensional grid graphs, and small world graphs.

A heuristic approach to dealing with dynamic scenarios, which is based on highway hierarchies, has been developed by Nannicini et al. [65].

Transit-Node Routing   is based on two key observations: First, there is a relatively small set of *transit nodes*—about 10 000 for the Western European or the US road network—with the property that for every pair of nodes that are 'not too close' to each other, the shortest path between them passes through at least one of these transit nodes. Second, for every node, the set of transit nodes encountered first when going far—so-called *access nodes*— is small. When distances from all nodes to their respective access nodes and between all transit nodes have been precomputed, a 'non-local' shortest-path query can be reduced to a few table lookups. An important ingredient is a *locality filter* that decides whether source and target are too close so that a special treatment is required to guarantee the correct result. In order to handle such local queries more efficiently, further levels can be added to the basic approach.

A generic framework for transit-node routing and a concrete instantiation based on highway hierarchies have been introduced in [71, 4, 5]. The generic framework is also a part of this thesis (Chapter 6), accompanied by an instantiation based on highway-node routing[6]. In addition to the already mentioned instantiations, there are two other implementations:

Separator-Based Transit-Node Routing. Using more space and pre-processing time, the separator-based multi-level method can be extended to implement transit-node routing: The separator nodes become transit nodes and the access nodes of v are the border nodes of the component of v. Local queries are those within a single component. Another level of transit nodes can be added by recursively finding separators of each component. Independently from our work, Müller et al. have essentially developed this approach, using different terminology[7]. Note that their first results [63] were published before any other implementation of transit-node routing. However, it took some time till reliable measurement data were available[8]

---

[6]Highway hierarchies (Chapter 3) and highway-node routing (Chapter 4) are two related route planning techniques. When we implemented transit-node routing for the first time, highway-node routing was not available yet. Therefore, the mentioned publications [71, 4, 5] are based on highway hierarchies. After highway-node routing has been developed, we reimplemented transit-node routing since by this means the preprocessing times could be considerably reduced. We decided to include the most recent version (based on highway-node routing) in this thesis.

[7]We chose to interpret their work using the transit-node terminology in order to point out similarities to our work.

[8]In their implementation, the preprocessed data is stored on a hard disk. Using a more compact representation, the data would fit into main memory. Therefore, when measuring query times, it is justifiable to assume that the required data was in main memory. This situation makes performing experiments more difficult.

[16]. An interesting difference to generic transit–node routing is that the re–quired information for routing between any pair of components is arranged together. This takes additional space but has the advantage that the informa–tion can be accessed more cache efficiently (it also allows subsequent space optimisations).

Although separators of road networks have much better properties than the worst case bounds for planar graphs would suggest, separator–based transit node routing needs about 4–8 times as many access nodes as our scheme (depending on the used metric) leading to much higher preprocess–ing times. The main reason for the difference in number of access nodes is that the separator approach does not take the 'sufficiently far away' criterion into account that is so important for reducing the number of access nodes in our implementations, in particular in case of the travel time metric.

Grid-Based Transit-Node Routing.   Bast, Funke and Matijevic pro–posed the transit–node routing approach based on a geometric grid [3]: The network is subdivided into uniform cells. Border nodes of these cells that are needed for 'long–distance' travel are used as access nodes. The union of all access nodes forms the transit–node set. As a locality filter it is sufficient to check whether source and target lie a certain number of cells apart.

They were the first to explicitly formulate the central observations and concepts of transit–node routing[9]. Our work was completed a few weeks later and has been accomplished largely independently from theirs except for the fact that their observation that about ten access nodes per node were sufficient motivated us to rethink our access node definition leading to a con–siderable reduction from around 55 to about ten, which made an implemen–tation for large graphs much more practicable, accelerated our development process significantly and yielded very good query times. While most algo–rithms described in [3] cater to the specific grid–based approach, we prefer a more generic notion of transit–node routing and regard our implementa–tions based on highway hierarchies and highway–node routing only as two possible (and very successful) instantiations of transit–node routing.

---

[9]In particular, they introduced the term 'transit node'. In a joint paper [4], we adopted some formulations and terms from [3] to describe the generic approach. For the sake of simplicity, we decided to keep these phrases in this thesis.

In a joint paper [4], the grid–based implementation and the one based on highway hierarchies are contrasted. One noticeable difference is that the variant based on highway hierarchies deals with all types of queries in a highly efficient way, while the grid–based variant only answers non–local queries very quickly (which, admittedly, constitute a very large fraction of all queries if source and target are picked uniformly at random). The grid–based variant is designed for comparatively modest memory require-ments, while our highway–hierarchy–based implementation has significantly smaller preprocessing and average query times. Note that our implementa-tion would need considerably less memory if we concentrated only on undi-rected graphs and non–local queries as it is done in the grid–based implemen-tation. Section 7.10.1 contains some concrete figures on the performance of grid–based transit–node routing.

1.2.4 Combinations

Many of the above techniques can be combined. In [79], a combination of a special kind of geometric container, the separator–based multi–level method, and A^* search yields a speedup of 62 for a railway transportation problem. In [39, 98], combinations of A^* search, bidirectional search, the multi–level method, and geometric containers are studied: Depending on the graph type, different combinations turn out to be best. For real–world graphs, a com-bination of bidirectional search and geometric containers leads to the best running times.

REAL.   Goldberg et al. [26, 30, 31] have successfully combined their ad-vanced version of REach–based routing with landmark–based A^* search (the ALt algorithm), obtaining the REAL algorithm. Its query performance is similar to our highway hierarchies, while the preprocessing times are usu-ally worse. In the most recent version [30, 31], they introduce a variant where landmark distances are stored only with the more important nodes, i.e., nodes with high reach values.[10] By this means, the memory consump-tion can be reduced significantly. Note that we developed a very similar idea independently when we combined highway hierarchies with the ALT algo-rithm [17] (Section 3.6). A comparison between REAL and our approaches is included in Section 7.10.1.

---

[10]They have already briefly mentioned this idea in [26].

SHARC [7] extends and combines ideas from highway hierarchies (namely, the contraction phase, which produces SHortcuts) with the edge flag (also called ARC flag) approach. The result is a fast *unidirectional* query algorithm, which is advantageous in scenarios where bidirectional search is prohibitive. In particular, using an approximative variant allows dealing with time–dependent networks efficiently. Even faster query times[11] can be obtained when a bidirectional variant is applied. The preprocessing times are slower than those of highway hierarchies, but considerably faster than those of the pure edge flag method.

Highway-Node Routing and Edge Flags. In his diploma thesis [73], Schieferdecker combines highway–node routing (Chapter 4) with the edge flag approach. Similarly to ideas used in other combinations of a hierar– chical with a goal–directed approach (e.g., REAL (see above) or HH* (Sec– tion 3.6)), preprocessing time and memory consumption can be kept low when the edge flags are computed not for the complete graph, but only for some level of the hierarchy. Query times of less than $100\,\mu s$ are obtained for the Western European road network. Schieferdecker also studies various other combinations, for example graph contraction with ALT or reach–based routing with edge flags.

1.2.5   Many-to-Many Shortest Paths

So far, the related–work section concentrated on the point–to–point prob– lem. In this subsection, we deal with the many–to–many variant of the shortest–path problem, where we want to compute an $|S| \times |T|$ distance table. As a naive solution, we can either solve $|S|$ single–source problems using Dijkstra's algorithm or we can employ *any* point–to–point speedup technique $|S| \times |T|$ times. There are results that accelerate many–to–many shortest paths for rather dense graphs with $m \gg n$ (e.g., [97]), which, how– ever, are not useful for road networks (or any other kind of sparse graphs). In his diploma thesis [50], Knopp adapted bidirectional search, geometric A^* search, and the ALT algorithm to the many–to–many case, yielding in the first two cases speedup factors up to 2 or 3 and in the third case factors up to 3 or 4 (depending on the type of input).

---

[11]We do not quote exact numbers since the final version of the paper is not ready yet.

1.2.6 Further Remarks

Some approaches (e.g. geometric containers) require for each node its geo–graphic coordinates, which might not always be available. However, there are studies that indicate that it is possible to *generate a layout* of a graph so that speedup techniques can be applied successfully. In some cases (where an original layout is available), generated layouts even result in a slightly higher speedup than the original layout does. [10, 11] deals with the special case of a timetable information system; a more general approach is presented in [95].

1.3 Main Contributions

1.3.1 Overview

We present three different point–to–point route planning techniques that compute provably optimal results. Our approaches exhibit various bene–fits and provide different kinds of trade–offs between preprocessing time, space consumption and query times. In addition, we introduce an algo–rithm that deals with the many–to–many variant. In an extensive experimen–tal study, we evaluate our algorithms using real–world road networks with up to 33 726 989 nodes. We do not only give average query times, but also detailed analyses of queries with different degrees of difficulty, per–instance worst–case upper bounds, and comparisons to other speedup techniques. For our standard test case, a network of Western Europe with about 18 mil–lion nodes, our lowest observed average query time is 4.3 μs on a 2.0 GHz machine, which corresponds to a speedup of 1.4 *million* compared to Dijk–stra's algorithm. Our fastest preprocessing time is 13 minutes and the lowest memory overhead is 0.7 bytes per node.[12] By setting a few tuning parame–ters appropriately, we can provide several good compromises between fast preprocessing, low memory consumption, and fast query times, which seem very reasonable for a wide range of practical applications. In selected cases, we also deal with a distance and a unit metric (instead of the usual travel time metric), turning restrictions, and outputting complete path descriptions.

---

[12]Note that these optima w.r.t. query time, preprocessing time, and memory consumption cannot be reached at the same time.

Moreover, we can handle dynamic scenarios: we can replace the en–tire cost function typically in less than two minutes or we can update only affected parts of the precomputed data structures if unexpected events like traffic jams occur: such an update operation takes about 40 ms in case of a single traffic jam on a motorway. Alternatively, we can even do without updating the data structures and instead perform a 'prudent' (and somewhat slower) query that takes the changed situation into account so that it still computes an optimal path.

Our many–to–many algorithm can compute a 10 000 times 10 000 dis–tance table in 23 seconds. Using Dijkstra's algorithm, the same task would take more than one day.

All algorithms that we present in this thesis are closely related. Fig–ure 1.4 gives an overview of the various relations.

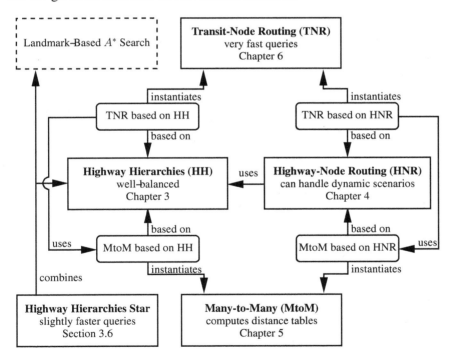

Figure 1.4: Overview of the relations between the route planning algorithms presented in this thesis. (Note that the landmark–based A^* search is *not* part of this thesis.)

1.3.2 Highway Hierarchies

The first version of highway hierarchies [75, 69] (Section 1.2.3) was a prototype, where we made several simplifying assumptions, in particular we dealt only with undirected graphs. While keeping the basic definition of the highway network and the idea to alternate between an edge reduction step (construction of the highway network) and a node reduction step (removing nodes of small degree), in this thesis, we greatly improve and generalise the highway hierarchies approach:

1. We fully support directed graphs. We even did some preliminary experiments where we also support turning restrictions.

2. The definition of the 'local area' has been generalised. Now, we can pick for each node an arbitrary individual neighbourhood radius that defines the neighbourhood (i.e., the local area) of that node.

3. We generalised the node reduction phase, extending the old trees-and-lines concept. The new method is not only simpler and directly applicable to directed graphs, but also more flexible and more effective. It has been successfully adopted by other route planning techniques, e.g. [31].

4. We give a simpler and more precise formulation of the query algorithm. It is very similar to bidirectional Dijkstra search with the difference that certain edges need not be expanded when the search is sufficiently far from source or target.

5. We present a complete proof of correctness that also rigorously handles the case that the shortest paths in the graph are not unique.

6. An all-pairs distance table for the topmost level L of the hierarchy is introduced. Forward and backward search can be stopped as soon as all entrance points to level L have been found. Then, the remaining gap can be bridged by performing a moderate number of simple table lookups. By this means, the query times are considerably improved. Furthermore, this optimisation can be seen as a pre-stage of transit-node routing (Section 1.2.3).

7. At the same time, we achieve an improvement w.r.t. preprocessing times, memory consumption, *and* query times. Using somewhat more memory, we obtain average query times below one millisecond on a 2.0 GHz machine—even for a road network with more than 30 million nodes.

8. We still cannot give a general worst-case bound better than Dijkstra's. So far, this drawback applies to all other exact speedup techniques where an implementation is available as well. However, in contrast to most of them, we can provide *per-instance worst-case guarantees*, i.e., we can give an upper bound for the search space size for *any* source-target query in a given graph without performing all n^2 possible queries, which would be prohibitive in a large road network. This is feasible since forward and backward search can be run independently from each other. Thus, we perform n forward and n backward searches and combine the observed search space sizes in an appropriate way.

9. The shortest paths that are determined by our query algorithm typically contain shortcut edges that have been created during the node reduction phase; each shortcut represents a path in the original graph. If we are interested not only in the shortest path length, but in the shortest path itself, we have to unpack the contained shortcuts, i.e., we have to determine the represented subpaths. We introduce space-efficient data structures that accomplish this task in a highly efficient manner.

Combination with Goal-Directed Search.   Since highway hierarchies lack any sense of goal direction, a combination with a goal-directed approach suggests itself. We went for a combination with landmark-based A^* search, which we call *highway hierarchies star*. Note that in case of highway hierarchies—in contrast to plain bidirectional Dijkstra or reach-based routing—, we are not allowed to abort the search as soon as forward and backward search meet. This fact turned out to be problematic for a combination with A^* search. Still, we managed to achieve a slight improvement w.r.t. query times. When using a distance metric (instead of the usual travel time metric) or when dealing with approximate queries, we even get considerable improvements. Furthermore, we introduce the idea to determine the landmarks not in the original graph, but in some level of the highway hierarchy. Since already the first node reduction phase typically leads to a network with less than one sixth of the original nodes, this accelerates the landmark selection considerably without observing a significant loss in quality of the selected landmark set. We can also compute and store the landmark distances only in some level k of the highway hierarchy, which reduces the memory consumption. When we use this optimisation, the query works in two phases: in an initial phase, we perform a non-goal-directed highway query until all entrance points to level k have been discovered; for the remaining search, the landmark distances are available so that we can use the combined algorithm in the main query phase. Note that Goldberg et al. [30, 31] have independently applied similar ideas to their combination of reach-based routing with landmark-based A^* search (Section 1.2.4).

1.3.3 Highway-Node Routing

Currently, highway hierarchies can handle only *static* road networks: when some edge weights change (e.g., due to a traffic jam), the precomputed hierarchy becomes obsolete. Since the procedure that constructs a highway network consists only of *local* searches, we anticipate that a highway hierarchy can be efficiently updated when edge weight changes occur since we need to repeat only the local searches that are potentially affected. Nevertheless, a proper realisation of this idea is nontrivial. One reason is that we would have to take care of two concepts, the edge reduction and the node reduction. Therefore, we aimed at the development of an even simpler route planning technique by factoring out some of the complications of highway hierarchies into a pre-preprocessing step. The result is an approach, called *highway-node routing*, that is useful both in static and in dynamic scenarios.

Highway-node routing is a generalisation of the separator-based multi-level method [37] (Section 1.2.3): we define overlay graphs using arbitrary node sets $V' \subseteq V$ rather than separators. New preprocessing and query algorithms are required since removing V' will in general *not* partition the graph into small components. To deal with this problem, we systematically investigate the graph theoretical problem of finding all nodes from V' that can be reached on a shortest path from a given node without passing another node from V'. The resulting algorithms form the crucial part of highway-node routing. The main remaining difficulty is to choose the highway nodes V'. The idea is that *important* nodes used by many shortest paths will lead to overlay graphs that are more sparse than for the separator-based approach. This will result in faster queries and low space consumption. The intuition behind this idea is that the number of overlay graph edges needed between the separator nodes bordering a region grows quadratically with the number of border nodes (see also [38]). In contrast, important nodes are uniformly distributed over the network and connected to only a small number of nearby important nodes.[13] While there are many ways to choose important nodes, we capitalise on previous results and use highway hierarchies to define the required node sets.[14]

Our method is so far the most space–efficient speedup technique that allows query times several thousand times faster than Dijkstra's algorithm. A direct comparison to the separator–based variant is difficult since previous papers use comparatively small graphs[15] and it is not clear how the original approach scales to very large graphs. Compared to highway hierarchies, highway–node routing has only slightly higher preprocessing times and similar query times.

Note that highway–node routing is conceptually simpler than highway hierarchies. In particular, we have only one construction step (overlay graph construction) instead of two (edge reduction and node reduction). This greatly simplifies dealing with *dynamic scenarios*. The idea is that in practice, a set of nodes important for one weight function will also contain most of the important nodes for another 'reasonable' weight function. The advantage is obvious when the cost function is redefined: all we have to do is to recompute the edge sets of the overlay graphs, which is by far faster than recomputing the underlying highway hierarchy. We also discuss two variants of the scenario when a few edge weights change: In a server setting, the affected parts of the overlay graphs are updated so that afterwards the static query algorithm will again yield correct results. In a mobile setting, the data structures are not updated. Rather, the query algorithm searches at lower levels of the node hierarchy, (only) where the information at the higher levels might have been compromised by the changed edges.

Together with [18], we were the first to present an approach that tackles such dynamic scenarios and to demonstrate its efficiency in an extensive experimental study using a real–world road network.

---

[13]This observation is also relevant for transit–node routing (cp. Section 1.2.3).

[14]Thus, the construction of a highway hierarchy constitutes a *pre-preprocessing* step of highway–node routing.

[15]For a subgraph of the European road network with about 100 000 nodes, [38] gives a preprocessing time of "well over half an hour [plus] several minutes" and a query time 22 times faster than Dijkstra's algorithm. For a comparison, we take a subgraph around Karlsruhe of a very similar size, which we preprocess in seven seconds. Then, we obtain a speedup of 94.

1.3.4 Many-to-Many Shortest Paths

In order to solve the many-to-many shortest-path problem, we present a generic algorithm that can be instantiated on the basis of any bidirectional and non-goal-directed point-to-point algorithm. The main idea is to perform only $|S|+|T|$ unidirectional queries instead of $|S| \times |T|$ bidirectional queries in order to compute an $|S| \times |T|$ distance table. Basically, this is done in the following way: We associate with each node in the graph a bucket that can store the distances to all reachable targets (and the node IDs of the corresponding targets). We perform $|T|$ backward searches: during a backward search from $t \in T$, we store the distance to t at each visited node. Then, we perform for each node $s \in S$ a forward search: at each visited node u, we scan its bucket and for each entry, we sum up the just computed distance from s to u and the stored distance from u to t; if applicable, we use the resulting sum to improve the minimum distance from s to t computed so far.

In order to get an efficient approach, we instantiate the generic algorithm using highway hierarchies and highway-node routing. Moreover, we introduce several optimisations; in particular, it turns out that a considerable asymmetry between forward and backward search is useful: we can accept larger forward search spaces if we can, in exchange, reduce the backward search space sizes because this decreases the number of bucket entries, which is advantageous since bucket scanning can become the bottleneck for large distance tables.

1.3.5 Transit-Node Routing

As already mentioned in Section 1.2.3, the central ideas of transit-node routing appear in three different realisations that have been developed largely (but not completely) independently of each other. Our starting point has been our highway hierarchies enhanced with an all-pairs distance table for the topmost level (Section 1.3.2). Sufficiently long shortest paths are composed of three parts: from the source to the forward entrance point to the topmost level, from the forward to the backward entrance point, and from the backward entrance point to the target. The distances of the first and the third part are computed during the query, the distance of the second part is looked up in the distance table. The essential step that leads to transit-node routing is to precompute also the distances of the first and the third part. We had to tackle three problems that remained:

1. We need to decide whether we can use the precomputed distances. If source and target are too close, we have to fall back to some special treatment, e.g., a local query. We implement such a *locality filter* using geometric disks that have the property that the disks of source and target overlap if the nodes are too close.

2. Originally, when we constructed a highway hierarchy of our Western European road network with a topmost level consisting of around 10 000 nodes, we observed about 55 entrance points to the topmost level. Although we consider this as an already quite small number, storing the distances from each node to all entrance points would have required sophisticated compression techniques to obtain data structures that still fit into main memory. Motivated by the results by Bast et al. [3], we realised that it is possible to jump to the topmost level earlier yielding fewer 'entrance points' (around 10), which we call 'access nodes' to distinguish them from the original entrance points. The reduced memory requirements simplified the implementation of transit–node routing based on highway hierarchies.

3. The redefinition of the access nodes requires that the all–pairs distance table contains the correct distances w.r.t. the original graph and not only w.r.t. the topmost level.[16] Computing the desired distance table can be done efficiently using our many–to–many algorithm (Section 1.3.4).

In addition to our concrete realisation based on highway hierarchies, one of our main contributions is to formulate a *generic framework* for transit–node routing that covers all existing implementations and that can be used as starting point for future instantiations. In particular, our framework extends transit–node routing to a *hierarchical approach* that consists of several levels: each level can have its own access nodes, an (only partly filled) distance table, and a locality filter. This way, all types of queries can be answered very efficiently.

---

[16]Note that in contrast to highway–node routing, a level of a highway hierarchy is not necessarily an overlay graph, i.e., it is not guaranteed that all distances in a highway network agree with the corresponding distances in the original graph.

After highway-node routing has been developed, we added a new instantiation of transit-node routing based on highway-node routing. It provides considerably smaller preprocessing times. Since both instantiations are conceptually very similar, in this thesis, we concentrate on the realisation based on highway-node routing since its performance is superior.

At the 9th DIMACS Implementation Challenge [1], our implementation based on highway hierarchies was the fastest participating speedup technique. The current version presented in this thesis is even slightly faster.

1.4 Outline

In Chapter 2, we present basic concepts from graph theory, priority queues, and Dijkstra's algorithm. The terminology and notation introduced in that chapter will be used throughout this thesis.

The arrangement of the main chapters of this thesis (Chapters 3–6) reflects the dependencies that are shown in Figure 1.4. We start with *highway hierarchies* (Chapter 3) that all other methods (more or less) rely on. In order to be self-contained, we give a complete account that also covers parts that have already been included in [75] and, thus, are not an official part of this thesis due to formal reasons.

Highway-node routing is presented in Chapter 4, followed by the *many-to-many* approach (Chapter 5). *Transit-node routing* somehow employs all other techniques. It is presented in Chapter 6. In spite of the existing dependencies, the main chapters are written in such a way that they can be read largely independently of each other.

In Chapter 7, we present an extensive experimental study that covers all route planning techniques introduced in this thesis. We also include some remarks on the implementation—more details concerning the implementation can be found in Appendix A.

This thesis is concluded in Chapter 8, which also contains some notes on possible future work.

2

Preliminaries

In this chapter, we introduce basic data structures, algorithms, and some notation that is used throughout this thesis. The presented concepts are covered
in more detail by virtually any textbook on algorithms, e.g. [14, 83].

2.1 Graphs and Paths

We expect a *directed* graph $G = (V, E)$ with a node set V of size n and an
edge set $E \subseteq V \times V$ of size m as input. A weight function $w : E \to \mathbb{R}_0^+$
assigns a *nonnegative* weight $w((u, v))$ to each edge (u, v). We usually just
write $w(u, v)$ instead of $w((u, v))$.

A *path* P in G from a node u_1 to a node u_k is a sequence of edges
$(u_1, u_2), (u_2, u_3), \ldots, (u_{k-1}, u_k)$. We often interpret such a path as a node
sequence $\langle u_1, u_2, \ldots, u_k \rangle$ or as a node set $\{u_1, u_2, \ldots, u_k\}$ if this simplifies
the notation. The *length* $w(P)$ of a path P is the sum of the weights of the
edges that belong to P. $P^* = \langle s, \ldots, t \rangle$ is a *shortest path* if there is no path
P' from s to t such that $w(P') < w(P^*)$. The *distance* $d_G(s, t)$ from s to
t in G is the length of a shortest path from s to t or ∞ if there is no path
from s to t. We just write $d(s, t)$ instead of $d_G(s, t)$ if G is clear from the
context. If $P = \langle s, \ldots, s', u_1, u_2, \ldots, u_k, t', \ldots, t \rangle$ is a path from s to t,
then $P|_{s' \to t'} = \langle s', u_1, u_2, \ldots, u_k, t' \rangle$ denotes the *subpath* of P from s' to
t'. We use $u \prec_P v$ to denote that a node u precedes[1] a node v on a path

---

[1]This does *not* necessarily mean that u is the *direct* predecessor of v.

$P = \langle \ldots, u, \ldots, v, \ldots \rangle$; we just write $u \prec v$ if the path P that is referred to is clear from the context. An example for these concepts is given in Fig. 2.1.

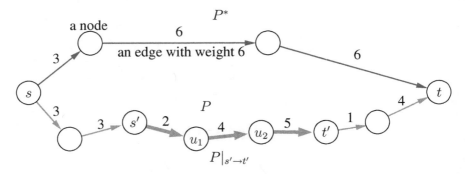

Figure 2.1: A directed graph with $n = 10$ nodes and $m = 10$ edges. There are two paths P and P^* from node s to node t. The length $w(P)$ of P is 22; $w(P^*) = 15$. P^* is a shortest path. The distance from s to t is $d(s, t) = w(P^*) = 15$. The edges of the subpath $P|_{s' \to t'}$ of P from s' to t' are represented by thick arrows. s' precedes u_2 on P, i.e., $s' \prec_P u_2$.

2.2   Priority Queues

A *priority queue* Q manages a set of elements with associated totally ordered priorities and supports the following operations:

- *insert* – insert an element into the priority queue,
- *deleteMin* – retrieve the element with the smallest priority and remove it from the priority queue,
- *decreaseKey* – set the priority of an element that already belongs to the priority queue to a new value that is less than the old value.

There are various ways to implement a priority queue (see also Section 1.2.1), for example using a simple *binary heap* that supports all operations in $O(\log(|Q|))$ or a considerably more complicated *Fibonacci heap* [24] that supports *insert* and *decreaseKey* in constant time and *deleteMin* in logarithmic amortised time.

2.3 Dijkstra's Algorithm

Single-Source Shortest-Path Problem. Dijkstra's algorithm [20] can be used to solve the *single-source shortest-path (SSSP) problem*, i.e., to com– pute the shortest paths from a single source node s to all other nodes in a given graph. Starting with the source node s as root, Dijkstra's algorithm grows a *shortest-path tree*[2] that contains shortest paths from s to all other nodes. During this process, each node of the graph is *unreached*, *reached*, or *settled*. A node that already belongs to the tree is *settled*. If a node u is settled, a shortest path P^* from s to u has been found and the distance $d(s, u) = w(P^*)$ is known. A node that is adjacent to a settled node is *reached*. Note that a settled node is also reached. If a node u is reached, a path P from s to u, which might not be the shortest one, has been found and a *tentative distance* $\delta(u) = w(P)$ is known. A node u that is not reached is *unreached*; for such a node, we have $\delta(u) = \infty$. The nodes that are reached but not settled are managed in a priority queue. The priority of a node u in the priority queue is u's tentative distance $\delta(u)$. Reached but not settled nodes are also called *queued*.

    Initially, s is inserted into the priority queue with the tentative distance 0. Thus, s is reached, all other nodes are unreached. While the priority queue is not empty, the node u with the smallest tentative distance is removed (*deleteMin*) and added to the shortest–path tree, i.e., u becomes settled. Fur– thermore, u's outgoing edges are *relaxed*:

- if an edge (u, v) leads to an unreached node v, v is added to the priority queue (*insert*); now, v is reached;

- if an edge (u, v) leads to a queued node v, v's key in the priority queue is updated (*decreaseKey*) provided that the length of the path from s via u to v is less than v's old key;

- if an edge (u, v) leads to a settled node v, it is ignored.

In case that the shortest paths in a graph are not unique, Dijkstra's algo– rithm can be easily modified to determine *all* shortest paths between s and any node $u \in V$. This means that not a shortest–path tree is grown, but a shortest–path *directed acyclic graph* (DAG).

---

[2]When we consider variants of Dijkstra's algorithm that are no longer guaranteed to only find shortest paths, we use the term *search tree* to denote the tree that the algorithm grows.

Dijkstra's algorithm involves n *insert*, n *deleteMin*, and at most m *decreaseKey* operations, yielding a runtime complexity of $O(m + n \log n)$ if, for example, Fibonacci heaps are used.

Point-to-Point Queries. If we are interested only in the shortest path(s) from the source node s to a single target node t, Dijkstra's algorithm can be stopped as soon as t has been settled.

A *bidirectional* version of Dijkstra's algorithm can be used to accelerate a shortest–path query from a given node s to a given node t. Two Dijkstra searches are executed in parallel: one searches from the source node s in the original graph $G = (V, E)$, also called *forward graph* and denoted as $\overrightarrow{G} = (V, \overrightarrow{E})$; another searches from the target node t backwards, i.e., it searches in the *reverse graph* $\overleftarrow{G} = (V, \overleftarrow{E})$, $\overleftarrow{E} := \{(v, u) \mid (u, v) \in E\}$. The reverse graph $\overleftarrow{G}$ is also called *backward graph*. When both search scopes meet, a shortest path from s to t can be easily derived by considering all elements that are currently queued.

Dijkstra Rank. Let us fix any rule that decides which element Dijkstra's algorithm removes from the priority queue in the case that there is more than one queued element with the smallest key. Then, during a Dijkstra search from a given node u, all nodes are settled in a fixed order. The *Dijkstra rank* $\text{rk}_u(v)$ of a node v is the rank of v w.r.t. this order. u has Dijkstra rank $\text{rk}_u(u) = 0$, the closest neighbour v_1 of u has Dijkstra rank $\text{rk}_u(v_1) = 1$, and so on.

3

Highway Hierarchies

3.1 Central Ideas

Let us consider the following naive route planning method:

1. Look for the next reasonable motorway.

2. Drive on motorways to a location close to the target.

3. Leave the motorway and search the target starting from the motorway exit.

Of course, it is true that this fast method does not always yield the optimal solution, but, in many cases, we obtain a reasonable approximation (provided that source and target are not too close together and that we travel in a country whose motorway network is well developed). This naive route planning method is based on a simple rule of thumb: when we are on our way to a remote target and pass by a city on a motorway, it usually does not pay to leave the motorway and look for a faster way through the city; in other words, usually, we can safely ignore all 'less important' city streets and stick to the 'more important' motorway since we *know* that the motorway provides the fastest way. The approach that is used by some commercial route planning systems is based on the above idea:

1. Search from the source and target node (*'bidirectional'*) within a certain radius (e.g. 20 km), consider *all roads*.

2. Continue the search within a larger radius (e.g. 100 km), consider only *national roads and motorways*.

3. Continue the search, consider only *motorways*.

Note that the actual implementations of this approach are more sophisticated than our simplified presentation suggests. Again, we get a method which is fast, but still returns inaccurate results—albeit better ones than those of the naive route planning method. We cannot guarantee exact results because we cannot exclude that sometimes it actually might be better to leave a 'more important' road (e.g. a motorway) and use some 'less important' street (e.g. a local road) that provides some kind of shortcut. In other words, a street that we considered to be 'less important' might turn out to be 'more important' than its category suggests. This observation is the starting point of our approach.

Similar to the commercial approach, we first perform a search in some *local area* around source and target; then, we switch to searching in a *highway network* that is much thinner than the complete graph. The crucial distinction of our approach is the fact that we define the notion of *local area* and *highway network* appropriately so that *exact* shortest paths can be computed. This is quite simple. For each node u, we set some *neighbourhood radius* and we define the neighbourhood of u (i.e., the local area around u) to consist of all nodes whose shortest-path distance from u does not exceed the neighbourhood radius. In our experiments, we do not use the same neighbourhood radius for each node, but we determine for each node its individual neighbourhood radius so that each neighbourhood contains the H closest nodes, where H is a tuning parameter. This is reasonable since road networks typically are quite heterogeneous: it would hardly be possible to pick a fixed neighbourhood radius that is suitable for both the city centre of Berlin and a rural area in Norway.

Our objective to obtain an exact algorithm requires the following definition of the highway network: An edge $(u, v) \in E$ belongs to the highway network if there are nodes s and t such that (u, v) is on some shortest path from s to t and not entirely within the neighbourhood of s or t. When we recall the intended query algorithm and consider the example in Figure 3.1, it gets obvious that this definition makes sense: During the forward search in the local area around s, we reach the node u; during the corresponding backward search, we reach v. Then, the search is continued only in the highway network. Thus, in order to guarantee that the shortest path can be found, all edges between u and v must belong to the highway network.

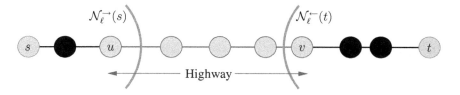

Figure 3.1: A shortest path from a node s to a node t. Edges that are not entirely within the neighbourhood of s or t are highway edges.

At first glance it might appear that a (prohibitively expensive) all-pairs shortest-path computation is needed to find the highway network. However, we will see that each highway edge is also within some local shortest-path tree B rooted at some $s \in V$ such that all leaves of B are 'sufficiently far away' from s.

Typically, a highway network contains a lot of nodes of small degree. For example, consider a motorway, which consists of a lot of road segments. The motorway is usually more important than the associated access ramps so that only the motorway might belong to the highway network, constituting a path of degree-2 nodes. In order to reduce the number of nodes, we contract the highway network by *bypassing* nodes with a small degree, introducing new shortcut edges, as illustrated in Figure 3.2. The result is a *contracted highway network*, also called *core*.

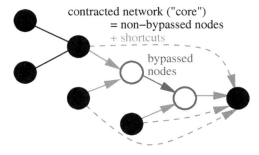

Figure 3.2: The core of a highway network consists of the subgraph induced by the set of non-bypassed nodes (solid) and additional shortcut edges (dashed).

Both concepts, the construction (which we also call *edge reduction*) and the contraction (which we also call *node reduction*) of the highway network, can be iterated, i.e., on the first core, we define local areas and construct a second highway network, contract it to obtain the second core, and so on. We arrive at a multi-level highway network—a *highway hierarchy*. The example in Table 3.1 illustrates the interaction of edge and node reduction. Each node reduction step increases the average node degree, while an edge reduction step decreases it again. All in all, the degree tends to go up from level to level, but the growth rate is very small. It is important to note that the shrinking factors do not change significantly from level to level (except for the very first and last level, perhaps).

Table 3.1: Construction of a highway hierarchy of the Western European road network with neighbourhood size $H = 70$, starting with a node reduction step. Note that the edge counters also include edges that can be only used in a backward search.

reduction type	#nodes	shrink factor	#edges	shrink factor	average degree
	18 029 721		44 448 388		2.5
node	2 739 750	6.6	21 311 324	2.1	7.8
edge	1 672 200	1.6	5 376 800	4.0	3.2
node	327 493	5.1	3 766 415	1.4	11.5
edge	270 606	1.2	1 109 315	3.4	4.1
node	72 787	3.7	981 297	1.1	13.5
edge	58 008	1.3	248 142	4.0	4.3
node	14 791	3.9	212 427	1.2	14.4
edge	11 629	1.3	53 744	4.0	4.6
node	2 941	4.0	46 632	1.2	15.9
edge	2 452	1.2	12 340	3.8	5.0
node	647	3.8	10 844	1.1	16.8
edge	569	1.1	3 076	3.5	5.4
node	163	3.5	2 808	1.1	17.2
edge	160	1.0	798	3.5	5.0
node	31	5.2	574	1.4	18.5

The query algorithm basically works in the following way: first, perform a local search in the original graph (level 0); second, switch to the highway network (level 1) and perform a local search in the highway network; then, switch to the next level of the highway hierarchy, and so on. Figure 3.3 gives a real–world example.

Figure 3.3: Search space for a query from Limburg (a German city) to a location 100 km east of the source node. Source and target are marked by a circle. The thicker the line, the higher the search level. Note that edges representing long subpaths are not drawn as direct shortcuts, but by showing the actual geographic route taken.

3.2 Definition

A *highway hierarchy* of a graph G consists of several levels $G_0, G_1, G_2, \ldots, G_L$, where the number of levels $L + 1$ is given. We will provide an inductive definition of the levels:

- Base case (G'_0, G_0): level 0 ($G_0 = (V_0, E_0)$) corresponds to the original graph G; furthermore, we define $G'_0 := G_0$.

- First step ($G'_\ell \rightarrow G_{\ell+1}, 0 \leq \ell < L$): for given *neighbourhood radii*, we will define the *highway network* $G_{\ell+1}$ of a graph G'_ℓ.

- Second step $(G_\ell \to G'_\ell, 1 \leq \ell \leq L)$: for a given set $B_\ell \subseteq V_\ell$ of *bypass-able* nodes, we will define the *core* G'_ℓ of level ℓ.

First step (*highway network*). For each node u, we choose nonnegative *neighbourhood radii* $r_\ell^{\rightarrow}(u)$ and $r_\ell^{\leftarrow}(u)$ for the forward and backward graph, respectively. To avoid some case distinctions, we set $r_\ell^{\rightarrow}(u)$ and $r_\ell^{\leftarrow}(u)$ to infinity for $u \notin V'_\ell$ (Radius Property R1) and for $\ell = L$ (R2). In all other cases, neighbourhood radii have to be $\neq \infty$ (R3).

The level-ℓ *neighbourhood* of a node $u \in V'_\ell$ is $\mathcal{N}_\ell^{\rightarrow}(u) := \{v \in V'_\ell \mid d_\ell(u,v) \leq r_\ell^{\rightarrow}(u)\}$ with respect to the forward graph and, analogously, $\mathcal{N}_\ell^{\leftarrow}(u) := \{v \in V'_\ell \mid d_\ell^{\leftarrow}(u,v) \leq r_\ell^{\leftarrow}(u)\}$ with respect to the backward graph, where $d_\ell(u,v)$ denotes the distance from u to v in the forward graph G_ℓ and $d_\ell^{\leftarrow}(u,v) := d_\ell(v,u)$ in the backward graph $\overleftarrow{G_\ell}$.

The *highway network* $G_{\ell+1} = (V_{\ell+1}, E_{\ell+1})$ of a graph G'_ℓ is defined by the set $E_{\ell+1}$ of *highway edges*: an edge $(u,v) \in E'_\ell$ belongs to $E_{\ell+1}$ iff there are nodes $s, t \in V'_\ell$ such that the edge (u,v) appears in some shortest path $\langle s, \ldots, u, v, \ldots, t \rangle$ from s to t in G'_ℓ with the property that $v \notin \mathcal{N}_\ell^{\rightarrow}(s)$ and $u \notin \mathcal{N}_\ell^{\leftarrow}(t)$. The set $V_{\ell+1}$ is the maximal subset of V'_ℓ such that $G_{\ell+1}$ contains no isolated nodes.

Second step (*core*). For a given set $B_\ell \subseteq V_\ell$ of *bypassable* nodes, we define the set S_ℓ of *shortcut edges* that bypass the nodes in B_ℓ: for each path $P = \langle u, b_1, b_2, \ldots, b_k, v \rangle$ with $u, v \in V_\ell \backslash B_\ell$ and $b_i \in B_\ell, 1 \leq i \leq k$, the set S_ℓ contains an edge (u,v) with $w(u,v) = w(P)$. The *core* $G'_\ell = (V'_\ell, E'_\ell)$ of level ℓ is defined in the following way:

$$V'_\ell := V_\ell \backslash B_\ell \quad \text{and} \quad E'_\ell := (E_\ell \cap (V'_\ell \times V'_\ell)) \cup S_\ell.$$

Removing all core nodes from G_ℓ yields connected *components of bypassed nodes*.

The *level* $\ell(e)$ *of an edge* e is $\max\{\ell \mid e \in E_\ell \cup S_\ell\}$. For an edge (u,v), we usually write just $\ell(u,v)$ instead of $\ell((u,v))$. The highway hierarchy can be interpreted as a single graph $\mathcal{G} := (V, E \cup \bigcup_{i=1}^{L} S_i)$ where each node and each edge has additional information on its membership in the various sets $V_\ell, V'_\ell, B_\ell, E_\ell, E'_\ell, S_\ell$.

3.3 Construction

3.3.1 Computing the Highway Network

Neighbourhood Radii. We suggest the following strategy to set the neighbourhood radii. For this paragraph, we interpret the graph G'_ℓ as an undirected graph, i.e., a directed edge (u, v) is interpreted as an undirected edge $\{u, v\}$ even if the edge (v, u) does not exist in the directed graph. Let $d_\ell^{\leftrightarrow}(u, v)$ denote the distance between two nodes u and v in the undirected graph. For a given parameter H_ℓ, for any node $u \in V'_\ell$, we set $r_\ell^{\rightarrow}(u) := r_\ell^{\leftarrow}(u) := d_\ell^{\leftrightarrow}(u, v)$, where v is the node whose Dijkstra rank $\mathrm{rk}_u(v)$ (w.r.t. the undirected graph) is H_ℓ. For any node $u \notin V'_\ell$, we set $r_\ell^{\rightarrow}(u) := r_\ell^{\leftarrow}(u) := \infty$ (to fulfil R1).

Originally, we wanted to apply the above strategy to the forward and backward graph one after the other in order to define the forward and backward radius, respectively. However, it turned out that using the same value for both forward and backward radius yields a similar good performance, but needs only half the memory.

Fast Construction: Outline. Given a graph G'_ℓ, we want to construct a highway network $G_{\ell+1}$. We start with an empty set of highway edges $E_{\ell+1}$. For each node $s_0 \in V'_\ell$, two phases are performed: the forward construction of a partial shortest–path directed acyclic graph (DAG) B (containing *all* shortest paths from s_0 to any node $u \in B$) and the backward evaluation of B. The construction is done by an SSSP search from s_0; during the evaluation phase, paths from the leaves of B to the root s_0 are traversed and for each edge on these paths, it is decided whether to add it to $E_{\ell+1}$ or not. The crucial part is the specification of an abort criterion for the SSSP search in order to restrict it to a 'local search'.

Phase 1: Construction of a Partial Shortest-Path DAG. A Dijkstra search from s_0 is executed. In order to keep track of all shortest paths, for each node in the partial shortest–path DAG B, we manage a list of (tentative) parents: when an edge (u, v) is relaxed such that $d_\ell(s_0, u) + w(u, v) = \delta(v)$, then u is added to the list of tentative parents of v. During the search, a reached node is either in the state *active* or *passive*. The source node s_0 is active; each node that is reached for the first time (*insert*) and each reached node that is updated (*decreaseKey*) is set to active iff any of its tentative par–ents is active. When a node p is settled, we consider all shortest paths P' from s_0 to p as depicted in Figure 3.4. The state of p is set to passive if

$$\forall \text{ shortest paths } P' = \langle s_0, \ldots, p \rangle :$$

$$s_1 \prec p \wedge p \notin \mathcal{N}_\ell^{\rightarrow}(s_1) \wedge s_0 \notin \mathcal{N}_\ell^{\leftarrow}(p) \wedge |P' \cap \mathcal{N}_\ell^{\rightarrow}(s_1) \cap \mathcal{N}_\ell^{\leftarrow}(p)| \leq 1 \tag{3.1}$$

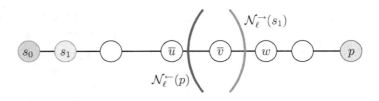

Figure 3.4: Abort criterion.

When no active unsettled node is left, the search is *aborted* and the growth of B stops.

An example for Phase 1 of the construction is given in Figure 3.5. The intuitive reason for s_1 (which is the first successor of s_0 on the path P') to appear in the abort criterion is the following: When we deactivate a node p during the search from s_0, we decide to ignore everything that lies behind p. We are free to do this because the abort criterion ensures that s_1 can take 'responsibility' for the things that lie behind p, i.e., further important edges will be added during the search from s_1. (Of course, s_1 will refer a part of its 'responsibility' to its successor, and so on.)

Phase 2: Selection of the Highway Edges. During Phase 2, exactly all edges (u, v) are added to $E_{\ell+1}$ that lie on paths $\langle s_0, \ldots, u, v, \ldots, p \rangle$ in the partial shortest–path DAG B with the property that $v \notin \mathcal{N}_\ell^{\rightarrow}(s_0)$ and $u \notin \mathcal{N}_\ell^{\leftarrow}(p)$. The example from Figure 3.5 is continued in Figure 3.6.

Theorem 1 *An edge $(u, v) \in E_\ell'$ is added to $E_{\ell+1}$ by the construction al-gorithm iff it belongs to some shortest path $P = \langle s, \ldots, u, v, \ldots, t \rangle$ and $v \notin \mathcal{N}_\ell^{\rightarrow}(s)$ and $u \notin \mathcal{N}_\ell^{\leftarrow}(t)$.*

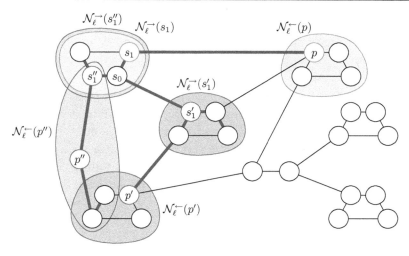

Figure 3.5: An example of Phase 1 of the construction. The weight of an edge is the length of the line segment that represents the edge in this figure. The neighbourhood size H_ℓ is 3. An SSSP search is performed from s_0. The abort criterion applies three times, at nodes p, p', and p''. All edges that belong to s_0's partial shortest-path tree are drawn as thick lines.

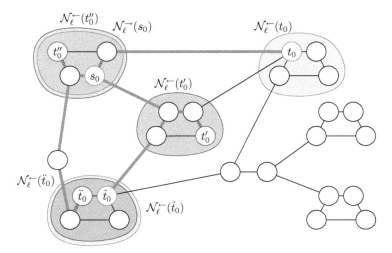

Figure 3.6: An example of Phase 2 of the construction. s_0's partial shortest path tree (thick lines) has five leaves t_0, t_0', t_0'', $\hat{t}_0$, and $\ddot{t}_0$. The edges that are added to $E_{\ell+1}$ are represented as solid thick lines.

Proof. In this proof, we will refer to the following Neighbourhood Prop‐erty N1 that follows directly from the neighbourhood definition: Consider a shortest path $\langle s, \ldots, u, \ldots, t \rangle$ in G'_ℓ. Then, $t \in \mathcal{N}_\ell^{\rightarrow}(s)$ implies $u \in \mathcal{N}_\ell^{\rightarrow}(s)$ and $s \in \mathcal{N}_\ell^{\leftarrow}(t)$ implies $u \in \mathcal{N}_\ell^{\leftarrow}(t)$.

$\Leftarrow$) Consider the node s_0 on $P|_{s \rightarrow u}$ such that $v \notin \mathcal{N}_\ell^{\rightarrow}(s_0)$ and $d_\ell(s_0, v)$ is minimal. Such a node s_0 exists because the condition $v \notin \mathcal{N}_\ell^{\rightarrow}(s_0)$ is always fulfilled for $s_0 = s$. The direct successor of s_0 on P is denoted by s_1. Note that $v \in \mathcal{N}_\ell^{\rightarrow}(s_1)$ [*]. We show that the edge (u, v) is added to $E_{\ell+1}$ when Phase 1 and 2 are executed from s_0. Due to the specification of Phase 2, it is sufficient to prove that after Phase 1 has been completed, the partial shortest‐path DAG B contains a node $p \in P|_{s_0 \rightarrow t}$ such that $v \preceq p$ and $u \notin \mathcal{N}_\ell^{\leftarrow}(p)$.

If $t \in B$, this statement is obviously fulfilled for $p := t$ since $v \preceq t$ and $u \notin \mathcal{N}_\ell^{\leftarrow}(t)$. Otherwise ($t \notin B$), the search is not continued from some node $t_0 \prec t$ on $P|_{s_0 \rightarrow t}$. We can conclude that t_0 is passive because, otherwise, its successor on $P|_{s_0 \rightarrow t}$ would adopt its active state and the search would not be aborted at that time. Since s_0 is active and t_0 is passive, either t_0 or one of its ancestors must have been switched from active to passive. Let p denote the first passive node on $P|_{s_0 \rightarrow t} = \langle s_0, s_1, \ldots, p, \ldots, t_0, \ldots, t \rangle$. Due to the definition of the abort condition, we have $s_1 \prec p \wedge p \notin \mathcal{N}_\ell^{\rightarrow}(s_1) \wedge s_0 \notin \mathcal{N}_\ell^{\leftarrow}(p) \wedge |P' \cap \mathcal{N}_\ell^{\rightarrow}(s_1) \cap \mathcal{N}_\ell^{\leftarrow}(p)| \leq 1$ [**], where $P' = P|_{s_0 \rightarrow p}$. The facts that $v \in \mathcal{N}_\ell^{\rightarrow}(s_1)$ [see *] and $p \notin \mathcal{N}_\ell^{\rightarrow}(s_1)$ [see **] imply $v \prec p$ due to N1. In order to obtain a contradiction, we assume $u \in \mathcal{N}_\ell^{\leftarrow}(p)$. Since $s_0 \notin \mathcal{N}_\ell^{\leftarrow}(p)$ [see **], this implies $s_0 \prec u$ by N1. Hence, $s_1 \preceq u$. Because $v \in \mathcal{N}_\ell^{\rightarrow}(s_1)$ [see *], we obtain $u \in \mathcal{N}_\ell^{\rightarrow}(s_1)$ due to N1. Similarly, we get $v \in \mathcal{N}_\ell^{\leftarrow}(p)$ since $v \prec p$ and $u \in \mathcal{N}_\ell^{\leftarrow}(p)$. Thus, $\{u, v\} \subseteq P' \cap \mathcal{N}_\ell^{\rightarrow}(s_1) \cap \mathcal{N}_\ell^{\leftarrow}(p)$. Therefore, $|P' \cap \mathcal{N}_\ell^{\rightarrow}(s_1) \cap \mathcal{N}_\ell^{\leftarrow}(p)| \geq 2$, which is a contradiction to [**]. We can conclude that $u \notin \mathcal{N}_\ell^{\leftarrow}(p)$.

$\Rightarrow$) Since each path $\langle s_0, \ldots, u, v, \ldots, p \rangle$ in B is a shortest path, the claim follows directly from the specification of Phase 2. $\qquad\square$

Algorithmic Details: Phase 1. For an efficient implementation, we keep track of a *border distance* $b(x)$ and a *reference distance* $a(x)$ for each node x in B. Along a path P' as depicted in Figure 3.4, we assign $b(x)$ the dis‐tance from the root to the border of the neighbourhood of s_1 as soon as s_1 is settled. This value is passed to all successors on the path, which allows to determine the first node w outside $\mathcal{N}_\ell^{\rightarrow}(s_1)$, i.e., its direct predecessor $\overline{v}$ is the last node inside $\mathcal{N}_\ell^{\rightarrow}(s_1)$. In order to fulfil the abort condition, we have to make sure that $\overline{v}$ is the only node on P' within $\mathcal{N}_\ell^{\rightarrow}(s_1) \cap \mathcal{N}_\ell^{\leftarrow}(p)$. There–fore, we want to check whether $\overline{v}$'s direct predecessor $\overline{u}$ belongs to $\mathcal{N}_\ell^{\leftarrow}(p)$. To allow an easy check, we determine, store, and propagate the reference distance from s_0 to $\overline{u}$ as soon as w is settled. Knowing the reference dis–tance $d_\ell(s_0, \overline{u})$, the current distance $d_\ell(s_0, p)$ and p's neighbourhood radius $r_\ell^{\leftarrow}(p)$, checking $\overline{u} \notin \mathcal{N}_\ell^{\leftarrow}(p)$ is then straightforward. If there are several shortest paths from s_0 to some node x, we determine appropriate maxima of the involved border and reference distances.

More formally, for any node x in B, $\wp(x)$ denotes the set of parent nodes in B. To avoid some case distinctions, we set $\wp(s_0) := \{s_0\}$, i.e., the root is its own parent. For the root s_0, we set $b(s_0) := 0$ and $a(s_0) := \infty$. For any other node $x \neq s_0$, we define $b'(x) := d_\ell(s_0, x) + r_\ell^{\rightarrow}(x)$ if $s_0 \in \wp(x)$, and 0, otherwise; $b(x) := \max(\{b'(x)\} \cup \{b(y) \mid y \in \wp(x)\})$; $a'(x) := \max\{a(y) \mid y \in \wp(x)\}$; and $a(x) := \max\{d_\ell(s_0, u) \mid y \in \wp(x) \wedge u \in \wp(y)\}$ if $a'(x) = \infty \wedge d_\ell(s_0, x) > b(x)$, and $a'(x)$, otherwise.

Then, we can easily check the following abort criterion at a settled node p:

$$a(p) + r_\ell^{\leftarrow}(p) < d_\ell(s_0, p) \tag{3.2}$$

Lemma 1 *(3.2) implies (3.1).*

Proof. We prove the contraposition "¬ (3.1) implies ¬ (3.2)", i.e., we as–sume that there is some shortest path P' from s_0 to p such that $p \preceq s_1 \vee p \in \mathcal{N}_\ell^{\rightarrow}(s_1) \vee s_0 \in \mathcal{N}_\ell^{\leftarrow}(p) \vee |P' \cap \mathcal{N}_\ell^{\rightarrow}(s_1) \cap \mathcal{N}_\ell^{\leftarrow}(p)| \geq 2$ and show that $a(p) + r_\ell^{\leftarrow}(p) \geq d_\ell(s_0, p)$.
Case 1: $p \preceq s_1$. If $p = s_0$, then $a(p) = \infty$, which yields ¬ (3.2). Otherwise $(p = s_1)$, $b(p) \geq d_\ell(s_0, p) + r_\ell^{\rightarrow}(p)$, $a'(p) = \infty$, and $a(p) = a'(p)$ since $d_\ell(s_0, p) \leq b(p)$, which implies ¬ (3.2).
Case 2: $s_1 \prec p \wedge p \in \mathcal{N}_\ell^{\rightarrow}(s_1)$. Due to N1 (see proof of Theorem 1), we have $\forall x, s_1 \preceq x \preceq p : x \in \mathcal{N}_\ell^{\rightarrow}(s_1)$. Hence, $\forall x : d_\ell(s_0, x) \leq d_\ell(s_0, s_1) + r_\ell^{\rightarrow}(s_1) \leq b(x)$. By an inductive proof, we can show that $a(p) = \infty$, which yields ¬ (3.2).
Case 3: $s_1 \prec p \wedge p \notin \mathcal{N}_\ell^{\rightarrow}(s_1) \wedge s_0 \in \mathcal{N}_\ell^{\leftarrow}(p)$. We have $d_\ell(s_0, p) \leq r_\ell^{\leftarrow}(p)$, which directly implies ¬ (3.2).

Case 4: $s_1 \prec p \wedge p \notin \mathcal{N}_\ell^\rightarrow(s_1) \wedge s_0 \notin \mathcal{N}_\ell^\leftarrow(p) \wedge |P' \cap \mathcal{N}_\ell^\rightarrow(s_1) \cap \mathcal{N}_\ell^\leftarrow(p)| \geq 2$. The assumption of Case 4 implies that there are two nodes $\overline{u}$ and $\overline{v}$, $s_1 \preceq \overline{u} \prec \overline{v} \preceq p$, that belong to $P' \cap \mathcal{N}_\ell^\rightarrow(s_1) \cap \mathcal{N}_\ell^\leftarrow(p)$. If $a(p) = \infty$, we directly have $\neg$ (3.2). Otherwise, there has to be some node w on P' such that $a'(w) = \infty \wedge d_\ell(s_0, w) > b(w)$. Obviously, $w \neq s_0$. Consider such a node w that maximises $d_\ell(s_0, w)$, i.e., for all nodes $x \succ w$ the above stated condition does not hold, which implies $a(x) = a'(x) \geq a(w)$. In particular, $a(p) \geq a(w)$. We have $b(w) \geq d_\ell(s_0, s_1) + r_\ell^\rightarrow(s_1)$. We can conclude that $d_\ell(s_0, w) > d_\ell(s_0, s_1) + r_\ell^\rightarrow(s_1)$ and, thus, $w \notin \mathcal{N}_\ell^\rightarrow(s_1)$. We obtain, by N1, $\overline{u} \prec \overline{v} \prec w$. Hence, $a(w) \geq d_\ell(s_0, \overline{u})$, which implies $a(p) \geq d_\ell(s_0, \overline{u})$. Furthermore, since $\overline{u} \in \mathcal{N}_\ell^\leftarrow(p)$, we have $r_\ell^\leftarrow(p) \geq d_\ell(\overline{u}, p)$. Adding up the last two inequalities yields $a(p) + r_\ell^\leftarrow(p) \geq d_\ell(s_0, p)$, which corresponds to $\neg$ (3.2). □

Algorithmic Details: Phase 2. For a node $u \in B$, we define $\mathcal{B}(u) := \{u\} \cup \{v \mid v \text{ is a descendant of } u \text{ in } B\}$ and the *slack* $\Delta(u) := \min_{w \in \mathcal{B}(u)} (r_\ell^\leftarrow(w) - d_\ell(u, w))$. For a leaf b, we have $\mathcal{B}(b) = \{b\}$ and $\Delta(b) = r_\ell^\leftarrow(b)$. The slack of an inner node u can be computed using only the slacks of its children $\gamma(u)$: $\Delta(u) = \min(r_\ell^\leftarrow(u), \min_{c \in \gamma(u)} \Delta_c(u))$, where $\Delta_c(u) := \Delta(c) - d_\ell(u, c)$. This leads to an equivalent, recursive definition.

The tentative slacks $\widehat{\Delta}(u)$ of all nodes u in B are set to $r_\ell^\leftarrow(u)$. We process all nodes in the reverse order as they were settled. This guarantees that all descendants of some node u have been processed before u is processed. We can stop as soon as a node $u \in \mathcal{N}_\ell^\rightarrow(s_0)$ is encountered. We maintain the invariant that the tentative slack $\widehat{\Delta}(u)$ of an element u that is processed is equal to the actual slack $\Delta(u)$. When a node u is processed, for each parent p of u in B, we perform the following steps: compute $\Delta_u(p) = \Delta(u) - d_\ell(p, u)$; if $\Delta_u(p) < 0$, the edge (p, u) is added to $E_{\ell+1}$; if $\Delta_u(p) < \widehat{\Delta}(p)$, the tentative slack $\widehat{\Delta}(p)$ is set to $\Delta_u(p)$. Figure 3.7 gives an example.

Theorem 2 *An edge (u, v) is added to $E_{\ell+1}$ by the slack-based method introduced above iff it lies on a path $\langle s_0, \ldots, u, v, \ldots, p \rangle$ in the partial shortest-path DAG B with the property that $v \notin \mathcal{N}_\ell^\rightarrow(s_0)$ and $u \notin \mathcal{N}_\ell^\leftarrow(p)$.*

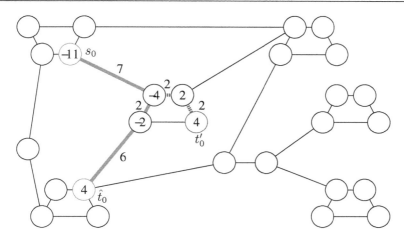

Figure 3.7: An example of the *slack-based method* that realises Phase 2 of the construction. The process is shown only for a part of the tree. As before, the weight of an edge is the length of the line that represents the edge in this figure. For the sake of transparency, the (rounded) weights are given explicitly for the relevant edges. Furthermore, the slacks of the involved nodes are given. Edges that are added to $E_{\ell+1}$ are solid, edges that are not added dotted.

Proof. $\Leftarrow$) From the definition of the slack of a node, it follows that

$$\begin{aligned} \Delta_v(u) &= \Delta(v) - d_\ell(u,v) \leq r_\ell^{\leftarrow}(p) - d_\ell(v,p) - d_\ell(u,v) \\ &= r_\ell^{\leftarrow}(p) - d_\ell(u,p) < 0 \end{aligned}$$

because $u \notin \mathcal{N}_\ell^{\leftarrow}(p)$. Since $v \notin \mathcal{N}_\ell^{\rightarrow}(s_0)$, v is processed at some point. Then, $\Delta_v(u)$ is computed and, since it is negative, the edge (u,v) is added to $E_{\ell+1}$.

$\Rightarrow$) Only edges that belong to a path in B from s_0 to a node p are con-sidered. The condition $v \notin \mathcal{N}_\ell^{\rightarrow}(s_0)$ is never violated because the traver-sal from the leaves to the root, and consequently, the addition of edges to $E_{\ell+1}$, is not continued when a node $v \in \mathcal{N}_\ell^{\rightarrow}(s_0)$ is encountered. If an edge (u,v) is added, the condition $\Delta_v(u) < 0$ is fulfilled. Hence, $\Delta(u) = \min_{w \in \mathcal{B}(u)} (r_\ell^{\leftarrow}(w) - d_\ell(u,w)) \leq \Delta_v(u) < 0$. Therefore, there is a node p such that $d_\ell(u,p) > r_\ell^{\leftarrow}(p)$, i.e., $u \notin \mathcal{N}_\ell^{\leftarrow}(p)$. $\qquad\square$

Theorem 3 *Let V_B denote the set of nodes of s_0's partial shortest-path DAG B. Let $G_B = (V_B, E_B)$ denote the subgraph of G'_ℓ that is vertex induced by V_B. The complexity of Phase 1 and 2 started from s_0 is $T_{Dijkstra}(|G_B|)$.*

Proof. The number of nodes of G_B is denoted by n', the number of edges by m'. The complexity of Phase 1 corresponds to the complexity of a SSSP search in G_B started from s_0, i.e., $O(n' + m')$ outside the priority queue plus n' *insert* and n' *deleteMin* operations plus at most m' *decreaseKey* operations. During Phase 2, each node and each edge is processed at most once, i.e., Phase 2 runs in $O(n' + m')$. □

Speeding Up the Highway Network Construction. Even a single active endpoint of a long edge (e.g., a long-distance ferry connection) can cause a large search space during construction, although most nodes of the search space might already be passive. To face this undesirable effect, we declare an active node v to be a *maverick* if $d_\ell(s_0, v) > f \cdot r_\ell^{\rightarrow}(s_0)$, where f is a parameter. When all active nodes are mavericks, the search from passive nodes is no longer continued. This way, the construction process is accelerated and $E_{\ell+1}$ becomes a superset of the highway network. Hence, queries will be slower, but still compute exact shortest paths. The *maverick factor* f enables us to adjust the trade-off between construction and query time.

3.3.2 Computing the Core

In order to obtain the core of a highway network, we contract it, which yields several advantages. The search space during the queries gets smaller since bypassed nodes are not touched and the construction process gets faster since the next iteration only deals with the nodes that have not been bypassed. Furthermore, a more effective contraction allows us to use smaller neighbourhood sizes without compromising the shrinking of the highway networks. This improves both construction and query times. However, bypassing nodes involves the creation of shortcuts, i.e., edges that represent the bypasses. Due to these shortcuts, the average degree of the graph is increased and the memory consumption grows. In particular, more edges have to be relaxed during the queries. Therefore, we have to carefully select nodes so that the benefits of bypassing them outweigh the drawbacks.

We give an iterative algorithm that combines the selection of the by-passable nodes B_ℓ with the creation of the corresponding shortcuts. We manage a stack that contains all nodes that have to be considered, initially all nodes from V_ℓ. As long as the stack is not empty, we deal with the topmost node u. We check the *bypassability criterion* #shortcuts $\leq$ $c \cdot (\deg_{in}(u) + \deg_{out}(u))$, which compares the number of shortcuts that would be created when u was bypassed with the sum of the in-and outdegree of u. The magnitude of the contraction is determined by the parameter c. If the criterion is fulfilled, the node is bypassed, i.e., it is added to B_ℓ and the appropriate shortcuts are created. Note that the creation of the shortcuts alters the degree of the corresponding endpoints so that bypassing one node can influence the bypassability criterion of another node. Therefore, all adjacent nodes that have been removed from the stack earlier, have not been bypassed yet, and are bypassable now are pushed on the stack once again.

Theorem 4 *If $c < 2$, $|E'_\ell|$ is in $O(|V_\ell| + |E_\ell|)$.*

Proof. If a node u is bypassed, the number of edges in the (tentative) core is increased by $\mathcal{D}_u := $ #shortcuts $- \deg_{in}(u) - \deg_{out}(u)$. (We have to subtract $\deg_{in}(u)$ and $\deg_{out}(u)$ since the edges incident to u no longer belong to the core.) Note that #shortcuts $= \deg_{in}(u) \cdot \deg_{out}(u) - \deg_{\leftrightarrow}(u)$, where $\deg_{\leftrightarrow}(u)$ denotes the number of adjacent nodes v that are connected to u by both an edge (u, v) and an edge (v, u). (We have to subtract $\deg_{\leftrightarrow}(u)$ to account for the fact that a 'shortcut' that would be a self-loop is not created.) We can conclude that $\mathcal{D}_u \leq \deg_{in}(u) \cdot \deg_{out}(u) - \deg_{in}(u) - \deg_{out}(u)$. If $\deg_{in}(u) \leq 1$ or $\deg_{out}(u) \leq 1$, we obtain $\mathcal{D}_u \leq 0$. Now, we deal with the case that $\deg_{in}(u) \geq 2$ and $\deg_{out}(u) \geq 2$. Since $\deg_{\leftrightarrow}(u) \leq \min(\deg_{in}(u), \deg_{out}(u))$, a node that fulfils the bypassability criterion also fulfils $\deg_{in}(u) \cdot \deg_{out}(u) \leq c \cdot (\deg_{in}(u) + \deg_{out}(u)) + \min(\deg_{in}(u), \deg_{out}(u))$. The inequality $x \cdot y \leq c(x + y) + \min(x, y)$ has only finitely many solutions (x, y) for $x, y \in \mathbb{N}, x, y \geq 2$ if $c \in \mathbb{R}$ is a constant less than 2. Consider the solution (x, y) that maximises $k := x \cdot y$. If there is no solution, take $k := 0$. Note that k is a constant that only depends on the constant c. We can conclude that $\mathcal{D}_u \leq k$.

Each node from V_ℓ is bypassed at most once. For each bypassed node, the number of edges in the (tentative) core is increased by at most k. Therefore, $|E'_\ell| \leq k \cdot |V_\ell| + |E_\ell|$. $\qquad\square$

If we used #shortcuts $\leq$ max$(\deg_{\text{in}}(u), \deg_{\text{out}}(u))$ as bypassability criterion, we would get a contraction that would be very similar to our earlier trees-and-lines method [69]. Note that the general version presented above allows a more effective contraction by setting c appropriately.

Limiting Component Sizes.   To reduce the observed maximum query time, we implement a limit on the number of hops a shortcut may repre‐ sent. By this means, the sizes of the components of bypassed nodes are reduced—in particular, the first contraction step tended to create quite large components of bypassed nodes so that it took a long time to leave such a component when the search was started from within it.

3.4   Query

Our *highway query algorithm* is a modification of the bidirectional version of Dijkstra's algorithm. Note that in contrast to the construction, during the query we need *not* keep track of ambiguous shortest paths. For now, we assume that the search is *not* aborted when both search scopes meet. This matter is dealt with in Section 3.4.2. We only describe the modifications of the forward search since forward and backward search are symmetric. In addition to the *distance* from the source, each node is associated with the search *level* and the *gap* to the 'next applicable neighbourhood border'. The search starts at the source node s in level 0. First, a local search in the neighbourhood of s is performed, i.e., the gap to the next border is set to the neighbourhood radius of s in level 0. When a node v is settled, it adopts the gap of its parent u minus the length of the edge (u, v). As long as we stay inside the current neighbourhood, everything works as usual. However, if an edge (u, v) crosses the neighbourhood border (i.e., the length of the edge is greater than the gap), we switch to a higher search level ℓ. The node u becomes an *entrance point* to the higher level. If the level of the edge (u, v) is less than the new search level ℓ, the edge is *not* relaxed—this is one of the two restrictions that cause the speedup in comparison to Dijkstra's algorithm (Restriction 1). Otherwise, the edge is relaxed: v adopts the new search level ℓ and the gap to the border of the neighbourhood of u in level ℓ since u is the corresponding entrance point to level ℓ. Figure 3.8 illustrates this process.

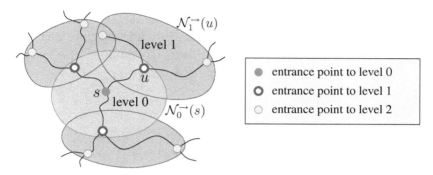

Figure 3.8: A schematic diagram of a highway query. Only the forward search is depicted.

We have to deal with the special case that an entrance point to level ℓ does not belong to the core of level ℓ. In this case, the search is continued inside a component of bypassed nodes till the level-ℓ core is entered, i.e., a node $u \in V'_\ell$ is settled. At this point, u is assigned the gap to the border of the level-ℓ neighbourhood of u. Note that before the core is entered (i.e., inside a component of bypassed nodes), the gap has been infinity (according to R1). To increase the speedup, we introduce another restriction (Restriction 2): when a node $u \in V'_\ell$ is settled, all edges (u, v) that lead to a bypassed node $v \in B_\ell$ in search level ℓ are *not* relaxed, i.e., once entered the core, we will never leave it again.

Figure 3.9 gives a detailed example of the forward search of a highway query. The search starts at node s. The gap of s is initialised to the distance from s to the border of the neighbourhood of s in level 0. Within the neighbourhood of s, the search process corresponds to a standard Dijkstra search. The edge that leads to u leaves the neighbourhood. It is not relaxed due to Restriction 1 since the edge belongs only to level 0. In contrast, the edge that leaves s_1 is relaxed since its level allows to switch to level 1 in the search process. s_1 and its direct successor are bypassed nodes in level 1. Their neighbourhoods are unbounded, i.e., their neighbourhood radii are infinity so that the gap is set to infinity as well. At s'_1, we leave the component of bypassed nodes and enter the core of level 1. Now, the search is continued in the core of level 1 within the neighbourhood of s'_1. The gap is set appropriately. Note that the edge to v is not relaxed due to Restriction 2 since v is a bypassed node. Instead, the direct shortcut to s_2 is used. Here, we switch to level 2. In this case, we do not enter the next level through a component of bypassed nodes, but we get directly into the core. The search is continued in the core of level 2 within the neighbourhood of s_2'. And so on.

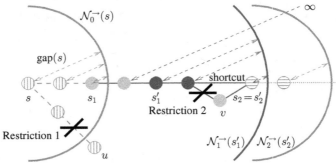

Figure 3.9: A detailed example of a highway query. Only the forward search is depicted. Nodes in level 0, 1, and 2 are vertically striped, solid, and horizontally striped, respectively. In level 1, dark shades represent core nodes, light shades bypassed nodes. Edges in level 0, 1, and 2 are dashed, solid, and dotted, respectively.

Despite of Restriction 1, we always find the optimal path since the construction of the highway hierarchy guarantees that the levels of the edges that belong to the optimal path are sufficiently high so that these edges are not skipped. Restriction 2 does not invalidate the correctness of the algorithm since we have introduced shortcuts that bypass the nodes that do not belong to the core. Hence, we can use these shortcuts instead of the original paths.

3.4.1 The Basic Algorithm

We use two priority queues $\overrightarrow{Q}$ and $\overleftarrow{Q}$, one for the forward search and one for the backward search. For each search direction, a node u is associated with a triple $(\delta(u), \ell(u), \mathrm{gap}(u))$, which we often call *key*. It consists of the (tentative) distance $\delta(u)$ from s (or t) to u, the search level $\ell(u)$, and the gap $\mathrm{gap}(u)$ to the next applicable neighbourhood border. Only the first component $\delta(u)$ is used to decide the priority within the queue.[1] We use

---

[1] If the search direction is not clear from the context, we will explicitly write $\overrightarrow{\delta}(u)$ and $\overleftarrow{\delta}(u)$ to distinguish between u's priority in $\overrightarrow{Q}$ and $\overleftarrow{Q}$.

the remaining two components for a tie breaking rule in the case that the same node is reached with two different keys $k := (\delta, \ell, \text{gap})$ and $k' := (\delta', \ell', \text{gap}')$ such that $\delta = \delta'$. Then, we prefer k to k' iff $\ell > \ell'$ or $\ell = \ell' \wedge \text{gap} < \text{gap}'$. Note that *any* other tie breaking rule (or even omitting an explicit rule) will yield a correct algorithm. However, the chosen rule is most aggressive and has a positive effect on the performance. Figure 3.10 contains the pseudo-code of the highway query algorithm.

input: source node s and target node t
output: distance $d(s,t)$

```
 1  d' := ∞;
 2  insert(Q⃗, s, (0, 0, r₀⃗(s))); insert(Q⃐, t, (0, 0, r₀⃐(t)));
 3  while (Q⃗ ∪ Q⃐ ≠ ∅) do {
 4      select direction ⇌ ∈ {→, ←} such that Q⇌ ≠ ∅;
 5      u := deleteMin(Q⇌);
 6      if u has been settled from both directions then
            d' := min(d', δ⃗(u) + δ⃐(u));
 7      if gap(u) ≠ ∞ then gap' := gap(u) else gap' := r⇌_{ℓ(u)}(u);
 8      foreach e = (u, v) ∈ E⇌ do {
 9          for (ℓ := ℓ(u), gap := gap';  w(e) > gap;
                ℓ++, gap := r⇌_ℓ(u));           // go "upwards"
10          if ℓ(e) < ℓ then continue;          // Restriction 1
11          if u ∈ V'_ℓ ∧ v ∈ B_ℓ then continue; // Restriction 2
12          k := (δ(u) + w(e), ℓ, gap − w(e));
13          if v has been reached then decreaseKey(Q⇌, v, k);
            else insert(Q⇌, v, k);
14      }
15  }
16  return d';
```

Figure 3.10: The highway query algorithm. Differences to the bidirectional version of Dijkstra's algorithm are marked: additional / modified lines have a framed line number; in modified lines, the modifications are underlined.

Remarks:

- Line 4: The correctness of the algorithm does not depend on the strategy that determines the order in which the forward and the backward searches are processed. However, the choice of the strategy can affect the running time in the case that an abort–on–success criterion is applied (see Section 3.4.2).

- Line 7: This line deals with the special case that the entrance point did not belong to the core when the current search level ℓ was entered, i.e., the gap was set to infinity. In this case, the gap is set to $r_{\overleftarrow{\ell(u)}}(u)$. This is correct even if u does not belong to the core, either, because in this case the gap stays at infinity.

- Line 9: It might be necessary to go upwards more than one level in a single step.

- Line 13: In the decreaseKey operation, the old key of v is only replaced by k if the above mentioned condition is fulfilled, i.e., if (a) the tentative distance improved or (b) stays unchanged while the tie breaking rule succeeds. In the latter case (b), no priority queue operation is invoked since the priority (the tentative distance) has not changed.[2]

The proof of correctness can be found in Section 3.5.

Algorithmic Details. If we group the outgoing edges (u, v) of each node u by level, we can avoid looking at edges (u, v) in levels $\ell(u, v) < \ell(u)$ since Restriction 1 would always apply to them. We can do without explicitly testing Restriction 2 if all edges (u, v) with $k := \ell(u, v), u \in V'_k$, and $v \in B_k$ have been downgraded to level $k - 1$. Then, the test of Restriction 1 also covers Restriction 2.

3.4.2 Optimisations

Rearranging Nodes. Similar to [31], after the construction has been completed, we rearrange the nodes by core level, which improves locality for the search in higher levels and, thus, reduces the number of cache misses.

---

[2]That way, we also avoid problems that otherwise could arise when an already settled node is reached once again via a zero weight edge.

Speeding Up the Search in the Topmost Level. Let us assume that we have a distance table that contains for any node pair $s, t \in V_L'$ the optimal distance $d_L(s, t)$. Such a table can be precomputed during the preprocessing phase by $|V_L'|$ SSSP searches in G_L'. Using the distance table, we do not have to search in level L. Instead, when we arrive at a node $u \in V_L'$ that leads to level L, we add u to the initially empty set $\overrightarrow{I}$ or $\overleftarrow{I}$ depending on the search direction; we do not relax the edge that leads to level L. After all entrance points have been encountered, we consider all pairs $(u, v) \in \overrightarrow{I} \times \overleftarrow{I}$ and compute the minimum path length $D := \overrightarrow{\delta}(u) + d_L(u, v) + \overleftarrow{\delta}(v)$. Then, the length of the shortest s-t-path is the minimum of D and the length d' of the tentative shortest path found so far (in case that the search scopes have already met in a level $< L$).

Using the distance table can be seen as extreme case of goal-directed search: from the nodes in the set $\overrightarrow{I}$, we directly 'jump' to the nodes in the set $\overleftarrow{I}$, which are close to the target. Thus, we can say that the highway query with the distance table optimisation works in two phases: a strictly non-goal-directed phase till the sets $\overrightarrow{I}$ and $\overleftarrow{I}$ have been determined, followed by a 'goal-directed jump' using the distance table.

For the sake of a simple incorporation of this optimisation into the high-way query algorithm, we slightly revise the properties R1 and R2: we use two distinguishable values ∞_1 and ∞_2 that are larger than any real number and set $r_\ell^{\rightleftharpoons}(u) := \infty_1$ for any ℓ and any node $u \notin V_\ell'$ (R1) and $r_L^{\rightleftharpoons}(u) := \infty_2$ for any node $u \in V_L'$ (R2). Then, we just add two lines to Figure 3.10 and modify Line 16:

between Lines 7 and 8:
7a **if** gap' $\neq \infty_1 \wedge \ell(u) = L$ **then** $\{ \overrightarrow{I} := \overrightarrow{I} \cup \{u\}; $ **continue;** $\}$

between Lines 11 and 12:
11a **if** gap $\neq \infty_1 \wedge \ell = L \wedge \ell > \ell(u)$ **then** $\{ \overrightarrow{I} := \overrightarrow{I} \cup \{u\}; $ **continue;** $\}$

16 **return** $\min(\{d'\} \cup \{ \overrightarrow{\delta}(u) + d_L(u, v) + \overleftarrow{\delta}(v) \mid u \in \overrightarrow{I}, v \in \overleftarrow{I} \});$

In Section 3.5.6, we show that our proof of correctness still holds when the distance table optimisation is applied.

Abort on Success. In the bidirectional version of Dijkstra's algorithm, we can abort the search as soon as both search scopes meet. Unfortunately, this would be incorrect for our highway query algorithm. The reason for this is illustrated in Figure 3.11. In the upper part of the figure, the bidirectional

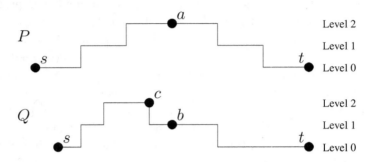

Figure 3.11: Schematic profile of a bidirectional highway query.

query from a node s to a node t along a path P is represented by a profile that shows the level transitions within the highway hierarchy. To get a sequential algorithm, at each iteration we have to decide whether a node from the forward or the backward queue is settled. We assume that a strategy is used that favours the smaller element. Thus, both search processes meet in the middle, at node a. When this happens, a path from s to t has been found. However, we have no guarantee that it is the shortest one. In fact, the lower part of the figure contains the profile of a shorter path Q from s to t, which is less symmetric than the profile of P. Note that the very flexible definition of the neighbourhoods allows such asymmetric profiles. When a on P is settled from both sides, b has been reached on Q by the backwards search, but *not* by the forward search since a search process never goes downwards in the hierarchy: therefore, at node c, the forward search is not continued on the path Q. We find the shorter path Q not until the backward search has reached c—which happens *after* P has been found. Hence, it would be wrong to abort the search, when a has been settled.

Therefore, we use a more conservative abort criterion: after a tentative shortest path P' has been encountered (i.e., after both search scopes have met), the forward (backward) search is not continued if the minimum element u of the forward (backward) queue has a key $\delta(u) \geq w(P')$. Obviously, the correctness of the algorithm is not invalidated by this abort criterion. In [69] we tried using more sophisticated criteria in order to reduce the search space. However, it turned out that this simple criterion, since it can be evaluated so efficiently, yields better query times in spite of a somewhat larger search space. Note that when the distance table optimisation is used and random queries are performed, our simple abort criterion is very close to an optimal criterion even with respect to the search space size: our experiments indicate that less than 1% of the search space is visited after the first meeting of forward and backward search.

3.4.3 Outputting Complete Path Descriptions

The highway query algorithm in Figure 3.10 only computes the distance from s to t. In order to determine the actual shortest path, we need to store pointers from each node to its parent in the search tree. Note that the algorithm could be easily modified to compute *all* shortest paths between s and t by just storing more than one parent pointer in case of ambiguities. However, subsequently, we only deal with a single shortest path.

We face two problems in order to determine a complete description of the shortest path: (a) we have to bridge the gap between the forward and backward topmost core entrance points (in case that the distance table optimisation is used) and (b) we have to expand the used shortcuts to obtain the corresponding subpaths in the original graph.

Problem (a) can be solved using a simple algorithm: We start with the forward core entrance point u. As long as the backward entrance point v has not been reached, we consider all outgoing edges (u, w) in the topmost core and check whether $d_L(u, w) + d_L(w, v) = d_L(u, v)$; we pick an edge (u, w) that fulfils the equation, and we set $u := w$. The check can be performed using the distance table. It allows us to greedily determine the next hop that leads to the backward entrance point.

Problem (b) can be solved without using any extra data (Variant 1). For each shortcut $(u, v) \in S_\ell$ on the shortest path, we perform a search from u to v in order to determine the represented path in G_ℓ. This search can be accelerated by using the knowledge that the first edge of the path enters a component C of bypassed nodes, the last edge leads to v, and all other edges are situated within the component C. The represented path in G_ℓ may contain shortcuts from sets $S_k, k < \ell$, which are expanded recursively. In the end, we obtain the represented path from u to v in the original graph.

However, if a fast output routine is required, it is necessary to spend some additional space to accelerate the unpacking process. We use a rather sophisticated data structure to represent unpacking information for the shortcuts in a space-efficient way (Variant 2). In particular, we do not store a sequence of node IDs that describe a path that corresponds to a short-cut, but we store only *hop indices*: for each edge (u, v) on the path that should be represented, we store its rank within the ordered group of edges that leave u. Since in most cases the degree of a node is very small, these hop indices can be stored using only a few bits. The unpacked shortcuts are stored in a recursive way, e.g., the description of a level-2 shortcut may contain several level-1 shortcuts. Accordingly, the unpacking procedure works recursively.

To obtain a further speed-up, we have a variant of the unpacking data structures (Variant 3) that caches the complete descriptions—without recursions—of all shortcuts that belong to the topmost level, i.e., for these important shortcuts that are frequently used, we do not have to use a recursive unpacking procedure, but we can just append the corresponding subpath to the resulting path.

3.4.4 Turning Restrictions

Real-world road networks can contain so-called *turning restrictions*. For example, a U-turn might be prohibited at certain traffic junctions. Formally, such a turning restriction (in its simplest and most common form) can be expressed as an edge pair $((u, v), (v, w))$: the edge (v, w) must not be traversed if the node v has been reached via the edge (u, v). Dealing with turning restrictions is a well-studied problem [74, 62]. In principle, there are two basic approaches: modifying the query algorithm or modelling the restrictions into the graph, which introduces additional artificial nodes and edges at affected road junctions. The latter technique can be applied irrespective of the used query algorithm.

In case of highway hierarchies, we expect that modelling turning restrictions into the graph only slightly deteriorates the performance since the artificial nodes usually have a very small degree so that most of them get bypassed in the very first contraction step. Furthermore, turning restrictions are often encountered at local streets that are not promoted to high levels of the hierarchy so that the negative impact is bounded to the lower levels. With respect to memory consumption, it is important to note that after the

preprocessing has been completed, artificial nodes and edges at road junctions that only belong to level 0 can be abandoned provided that the query
algorithm (which in level 0 just corresponds to Dijkstra's algorithm) is modified appropriately to handle turning restrictions.

3.5   Proof of Correctness

Difficulties.   Although the basic concepts (e.g. the definition of the highway network) and the algorithm are quite simple, the proof of correctness
gets surprisingly complicated. The main reason for that is the fact that we
cannot prove that *the* shortest path is found since there might be several
shortest paths of the same length. We could assume that the shortest paths
in the input are unique or that the uniqueness can be guaranteed by adding
small fractions to the edge weights as it is done by other authors who face
similar problems. However, the former would be too restrictive since usually, in real-world road networks, there are at least a few ambiguous instances, and a reliable realisation of the latter would be rather difficult. Furthermore, the introduction of shortcuts adds a lot of ambiguity even if it was
not present in the input.

   Therefore, if we pick any shortest path P to show that it is found by
the query algorithm, it can happen that a node u on P is settled from another node than its predecessor on P. Of course, in this case, u will still be
assigned the optimal distance from the source, but the search level and the
distance to the next neighbourhood border may be different than expected
so that we have to adapt to the new situation.

Outline.   We face the above mentioned difficulties in the following way:
First, we show that the algorithm terminates and deal with the special case
that no path from the source to the target exists (Section 3.5.1). Then, we
introduce some definitions and concepts that will be useful in the main part
of the correctness proof: In Section 3.5.2, we define for a given path, a corresponding *contracted* path and an *expanded* path, where subpaths in the
original graph are replaced by shortcuts or vice versa, respectively. In Section 3.5.3, we first define the concepts of *last neighbour* and *first core node*,
which, iteratively applied, lead to an *unidirectional labelling* of a given path.
Figure 3.12 gives an example. Applying a forward and a backward labelling

to the same path then allows the definition of a *meeting level* and a *meeting point* (Figure 3.13). The latter requires a case distinction since the forward and backward labelling may either meet in some core or in some compo-nent of bypassed nodes. Finally, we introduce the term *highway path*, a path whose properties exactly comply with the two restrictions of the query algorithm. Figure 3.14 gives an example.

In Section 3.5.4, we deal with the reachability along a highway path. Basically, we show that if the query has settled some node u on a highway path with the appropriate key, then u's successor on that path can be reached

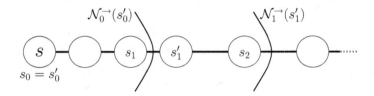

Figure 3.12: Example for a forward labelling of a path P. The labels s_0 and s_0' are set to s (base case). The node s_1 is the last neighbour of s_0' (denoted by $\overrightarrow{\omega}_0^P(s_0')$), the node s_1' is the first level-1 core node (denoted by $\overrightarrow{\alpha}_1^P(s_1)$), s_2 is the last neighbour of s_1', and so on.

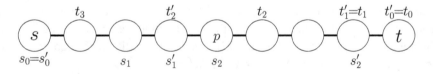

Figure 3.13: Example for a forward and backward labelling (depicted below and above the nodes, respectively). The meeting level is 2, the meeting point is p.

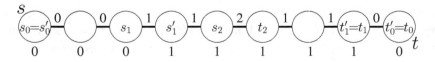

Figure 3.14: Example for a highway path. Each edge belongs at least to the given level, each node at least to the given core level.

from u with the appropriate key as well (Lemmas 6 and 7, which are proved using the auxiliary Lemma 5). In other words, if there is a highway path, the query can follow the path (at least if there was no ambiguity).

In Section 3.5.5, we use all concepts and lemmas introduced in the pre-ceding sections to conduct the actual correctness proof, where we also deal with ambiguous paths. The general idea is to say that at any point the query algorithm has some valid *state* consisting of a shortest s-t-path P and two nodes $u \preceq \bar{u}$ that split P into three parts such that the first and the third part are paths in the forward and backward search tree, respectively, and the second part is a contracted path. For such a valid state, we can prove that any node on the first and third part has been settled with the appropriate key (Lemma 8). Furthermore, we can show that P is a highway path (Lemma 9).

When the algorithm is started, the nodes s and t are settled and some shortest s-t-path P in the original graph exists. (The special case that no s-t-path exists has already been dealt with.) Consequently, our *initial* state is composed of the contracted version of P and the nodes s and t, which makes it a valid state. A *final* state is a valid state where forward and backward search have met, i.e., they have settled a common node $u = \bar{u}$. Originally, we wanted to show that a shortest path is found. Now, we see (in Lemma 10) that it is sufficient to prove that a final state is reached.

We have already defined the meeting point p on a path. We fall back on this definition and intend to prove that forward and backward search meet at p. When we look at any valid non–final state, it is obvious that at least one search direction can proceed to get closer to p, i.e., we have $u \prec p$ or $p \prec \bar{u}$ (Lemma 11). We pick such a *non-blocked* search direction. Let us assume w.l.o.g. that we picked the forward direction. We know that u has been settled with the appropriate key and that P is an optimal highway path (Lemmas 8 and 9). Due to the 'reachability along a highway path' (Lemmas 6 and 7), we can conclude that u's successor v can be reached with the appropriate key as well, in particular with the optimal distance from s. A node that can be reached with the optimal distance will also be settled at some point with the optimal distance. However, we cannot be sure that v is settled with u as its parent since the shortest path from s to v might be ambiguous. At this point the state concept gets handy: we just replace the subpath of P from s to v with the path in the search tree that actually has been taken, yielding a path P^+; we obtain a new state that consists of P^+

and the nodes v and $\overline{u}$. We prove that the new state is valid (Lemma 12).

Thus, we can show that from any valid non–final state another valid state is reached at some point. We also show in Lemma 12 that we cannot get into some cycle of states since in each step the length of the middle part of the path is decreased. Hence, starting from the initial state, eventually a final state is reached so that a shortest path is found (Theorem 5).

At this point, we will have proven the basic algorithm (Section 3.4.1) correct without considering the optimisations from Section 3.4.2. Finally, in Section 3.5.6, we show that the distance table optimisation does not in–validate the correctness. For the other two optimisations, this is obvious.

Additional Notations. '∘' denotes *path concatenation*. $\text{succ}(u, P)$ and $\text{pred}(u, P)$ denote the direct successor and predecessor of u on P, respec–tively. We just write $\text{succ}(u)$ and $\text{pred}(u)$ if the path is clear from the context. For two nodes u and v on some path, $\min(u, v)$ denotes u if $u \preceq v$ and v otherwise. $\max(u, v)$ is defined analogously. $d_P(u, v) := w(P|_{u \to v})$ denotes the distance from u to v along the path P. Note that for any edge (u, v) on P, we have $w(u, v) = d_P(u, v)$.

3.5.1 Termination and Special Cases

Since we have set the neighbourhood radius in the topmost level to infinity (R2), we are never tempted to go upwards beyond the topmost level. This observation is formalised in the following lemma.

Lemma 2 *The for-loop in Line 9 of the highway query algorithm always terminates with $\ell \leq L$ and $(\ell = L \to gap = \infty)$.*

Proof. We only consider iterations where the forward search direction is selected; analogous arguments apply to the backward direction. By an in–ductive proof, we show that at the beginning of any iteration of the main while–loop, we have $\ell(u) \leq L$ and $(\ell(u) = L \to gap(u) = \infty)$ for any node u in $\overrightarrow{Q}$.

Base Case: True for the first iteration, where only s belongs to $\overrightarrow{Q}$: we have $\ell(s) = 0 \leq L$ and $gap(s) = r_0^{\to}(s)$ (Line 2), which is equal to infinity if $L = 0$ (due to R2).

Induction Step: We assume that our claim is true for iteration i and show that it also holds for iteration $i + 1$. Due to the induction hypothesis, we have $\ell(u) \leq L$ and $(\ell(u) = L \rightarrow \text{gap}(u) = \infty)$ in Line 5. If $\ell(u) = L$, we have gap $= \text{gap}' = r_{\ell(u)}^{\rightarrow}(u) = \infty$ (Line 7 and 9, R2); thus the for-loop in Line 9 terminates immediately with $\ell = \ell(u) = L$. Otherwise ($\ell(u) < L$), the for-loop either terminates with $\ell < L$ or reaches $\ell = L$; in the latter case, we have gap $= r_{\ell}^{\rightarrow}(u) = \infty$ (Line 9, R2); hence, the loop terminates.

Thus, in any case, the loop terminates with $\ell \leq L$ and $(\ell = L \rightarrow$ gap $= \infty)$. Therefore, if the node v adopts the key k in Line 13 (either by a decreaseKey or an insert operation), the new key fulfills the required condition.

This concludes our inductive proof, which also shows that the claim of this lemma holds during any iteration of the main while-loop. $\qquad \square$

It is easy to the see that the following property of Dijkstra's algorithm also holds for the highway query algorithm.

Proposition 1 *For each search direction, the sequence of distances $\delta(u)$ of settled nodes u is monotonically increasing.*

Now, we can prove that

Lemma 3 *The highway query algorithm terminates.*

Proof. The for-loop in Line 9 always terminates due to Lemma 2. The for-loop in Line 8 terminates since the edge set is finite. The main while-loop in Line 3 terminates since each node v is inserted into each priority queue at most once, namely if it is unreached (Line 13); if it is reached, it either already belongs to the priority queue or it has already been settled; in the latter case, we know that $\delta(v) \leq \delta(u) \leq \delta(u) + w(e)$ (Proposition 1; edge weights are nonnegative) so that no priority queue operation is performed due to the specification of the decreaseKey operation. $\qquad \square$

The *special case* that there is no path from s to t is trivial. The algorithm terminates due to Lemma 3 and returns ∞ since no node can be settled from both search directions (otherwise, there would be some path from s to t). For the remaining proof, we assume that a shortest path from s to t exists in the original graph G.

3.5.2 Contracted and Expanded Paths

Lemma 4 *Shortcuts do not overlap, i.e., if there are four nodes $u \prec u' \prec v \prec v'$ on a path P in G, then there cannot exist both a shortcut (u, v) and a shortcut (u', v') at the same time.*

Proof. Let us assume that there is a shortcut $(u, v) \in S_\ell$ for some level ℓ. All inner nodes, in particular u', belong to B_ℓ. Since u' does not belong to V'_ℓ, a shortcut that starts from u' can belong only to some level $k < \ell$. If there was a shortcut $(u', v') \in S_k$, the inner node v would have to belong to B_k, which is a contradiction since $v \in V'_\ell$. $\square$

Definition 1 *For a given path P in a given highway hierarchy $\mathcal{G}$, the contracted path* $\mathrm{ctr}(P)$ *is defined in the following way: while there is a subpath $\langle u, b_1, b_2, \ldots, b_k, v \rangle$ with $u, v \in V'_\ell$ and $b_i \in B_\ell, 1 \leq i \leq k, k \geq 1$, for some level ℓ, replace it by the shortcut edge $(u, v) \in S_\ell$.*

Note that this definition terminates since the number of nodes in the path is reduced by at least one in each step and the definition is unambiguous due to Lemma 4.

Definition 2 *For a given path P in a given highway hierarchy $\mathcal{G}$, the level-ℓ expanded path* $\exp(P, \ell)$ *is defined in the following way: while the path contains a shortcut edge $(u, v) \in S_k$ for some $k > \ell$, replace it by the represented path in G_k.*

Note that this definition terminates since an expanded subpath can only contain shortcuts of a smaller level.

3.5.3 Highway Path

Consider a given highway hierarchy $\mathcal{G}$ and an arbitrary path $P = \langle s, \ldots, t \rangle$. In the following, we will bring out the structure of P w.r.t. $\mathcal{G}$.

Last Neighbour and First Core Node. For any level ℓ and any node u on P, we define the *last succeeding level-ℓ neighbour* $\overrightarrow{w}_\ell^P(u)$ and the *first succeeding level-ℓ core node* $\overrightarrow{\alpha}_\ell^P(u)$: $\overrightarrow{w}_\ell^P(u)$ is the node $v \in \{x \in P \mid u \preceq x \wedge d_P(u,x) \leq r_\ell^{\rightarrow}(u)\}$ that maximises $d_P(u,v)$, and $\overrightarrow{\alpha}_\ell^P(u)$ is the node $v \in \{t\} \cup \{x \in P \cap V_\ell' \mid u \preceq x\}$ that minimises $d_P(u,v)$. The *last preceding neighbour* $\overleftarrow{w}_\ell^P(u)$ and the *first preceding core node* $\overleftarrow{\alpha}_\ell^P(u)$ are defined analogously.

Unidirectional Labelling. Now, we inductively define a forward *labelling* of the path P. The labels s_0 and s_0' are set to s and for $\ell, 0 \leq \ell < L$, we set $s_{\ell+1} := \overrightarrow{w}_\ell^P(s_\ell')$ and $s_{\ell+1}' := \overrightarrow{\alpha}_{\ell+1}^P(s_{\ell+1})$. Furthermore, in order to avoid some case distinctions, $s_{L+1} := t$. For an example, we refer to Figure 3.12.

Proposition 2 *The following properties apply to the (Unidirectional) forward labelling of P:*

- U1: $s = s_0 = s_0' \preceq s_1 \preceq s_1' \preceq \ldots \preceq s_L \preceq s_L' \preceq s_{L+1} = t$
- U2a: $\forall \ell, 0 \leq \ell \leq L : \forall u, s_\ell' \preceq u \preceq s_{\ell+1} : d_P(s_\ell', u) \leq r_\ell^{\rightarrow}(s_\ell')$
- U2b: $\forall \ell, 0 \leq \ell \leq L : \forall u \succ s_{\ell+1} : d_P(s_\ell', u) > r_\ell^{\rightarrow}(s_\ell')$
- U3: $\forall \ell, 0 \leq \ell \leq L : \forall u, s_\ell \preceq u \prec s_\ell' : u \notin V_\ell'$
- U4: $\forall \ell, 0 \leq \ell \leq L : s_\ell' = t \vee s_\ell' \in V_\ell'$

A backward labelling (specifying nodes t_ℓ and t_ℓ') is defined analogously.

Meeting Level and Point. The *meeting level* λ of P is 0 if $s = t$ and $\max\{\ell \mid s_\ell \preceq t_\ell\}$ if $s \neq t$. Note that $\lambda \leq L$ (in any case) and $t_{\lambda+1} \prec s_{\lambda+1}$ (in case that $s \neq t$). The *meeting point* p of P is either t_λ (if $t_\lambda \preceq s_\lambda'$) or $\min(s_{\lambda+1}, t_\lambda')$ (otherwise). Figure 3.13 gives an example.

Proposition 3 *The following properties apply to the Meeting point p of P:*

- M1: $s_\lambda \preceq p \preceq t_\lambda$
- M2: $t_{\lambda+1} \preceq p \preceq s_{\lambda+1}$
- M3: $\forall \ell, 0 \leq \ell \leq L : (s_\ell' \prec p \rightarrow p \preceq t_\ell') \wedge (p \prec t_\ell' \rightarrow s_\ell' \preceq p)$

Proof. The case $s = t$ is trivial. Subsequently, we assume $s \neq t$. In order to prove M1, M2, and (M3 for $\ell = \lambda$), we distinguish between two cases.

Case 1: $t_\lambda \preceq s'_\lambda$. Then, $p = t_\lambda$. M1 is fulfilled due to the definition of the meeting level, which implies $s_\lambda \preceq t_\lambda$. Furthermore, due to U1, we have $t_{\lambda+1} \preceq t'_\lambda \preceq t_\lambda = p \preceq s'_\lambda \preceq s_{\lambda+1}$ so that M2 and (M3 for $\ell = \lambda$) are fulfilled.

Case 2: $s'_\lambda \prec t_\lambda$. Then, $p = \min(s_{\lambda+1}, t'_\lambda)$.

Subcase 2.1: $s_{\lambda+1} \preceq t'_\lambda$. Then, $p = s_{\lambda+1}$. We have $s_\lambda \preceq s'_\lambda \preceq s_{\lambda+1} = p \preceq t'_\lambda \preceq t_\lambda$ so that M1 and (M3 for $\ell = \lambda$) are fulfilled. Furthermore, M2 holds due to $t_{\lambda+1} \prec s_{\lambda+1}$.

Subcase 2.2: $t'_\lambda \prec s_{\lambda+1}$. Then, $p = t'_\lambda$. Since $s'_\lambda \prec t_\lambda \preceq t$, we know that $s'_\lambda \in V'_\lambda$ (due to U4). Thus, we have $s'_\lambda \preceq t'_\lambda \preceq t_\lambda$ (otherwise ($t'_\lambda \prec s'_\lambda \preceq t_\lambda$), we would have a contradiction with U3). Hence, $s_\lambda \preceq s'_\lambda \preceq t'_\lambda = p \preceq t_\lambda$ so that M1 and (M3 for $\ell = \lambda$) are fulfilled. M2 holds as well since $t_{\lambda+1} \preceq t'_\lambda = p \prec s_{\lambda+1}$.

It remains to show M3 for $\ell < \lambda$ and for $\ell > \lambda$. In the former case, M3 holds due to M1, which implies $s'_\ell \preceq s_\lambda \preceq p \preceq t_\lambda \preceq t'_\ell$ (U1). In the latter case, M3 holds due to M2, which implies $t'_\ell \preceq t_{\lambda+1} \preceq p \preceq s_{\lambda+1} \preceq s'_\ell$ (U1). $\square$

Highway Path. $P = \langle s, \ldots, t \rangle$ is a *highway path* (Figure 3.14) iff the following two <u>H</u>ighway properties are fulfilled:

- H1: $\forall \ell, 0 \le \ell \le L : \text{H1}(\ell)$
- H2: $\forall \ell, 0 \le \ell \le L : \text{H2}(\ell)$

where

- H1(ℓ): $\forall (u, v), s'_\ell \preceq u \prec v \preceq t'_\ell : u, v \in V'_\ell$
- H2(ℓ): $\forall (u, v), s_\ell \preceq u \prec v \preceq t_\ell : \ell(u, v) \ge \ell$

3.5.4 Reachability Along a Highway Path

We consider a path $P = \langle s, \ldots, t \rangle$. For a node u on P, we define the *reference level* $\overline{\ell}(u) := \max(\{0\} \cup \{i \mid s_i \prec u\})$.

Proposition 4 *For any two nodes u and v with $u \preceq v$, the following refer-ence <u>L</u>evel properties apply:*

- L1: $0 \leq \bar{\ell}(u) \leq L$
- L2: $s_{\bar{\ell}(u)} \preceq u$
- L3: $u \preceq s_{\bar{\ell}(u)+1}$
- L4: $\bar{\ell}(u) \leq \bar{\ell}(v)$

Definition 3 *A node u is said to be* <u>A</u>*ppropriately reached/settled with the key $k = (\delta(u), \ell(u), gap(u))$ on the path P iff all of the following conditions are fulfilled:*

- $A_1(k, u)$: $\delta(u) = d_0(s, u)$
- $A_2(k, u)$: $\ell(u) = \bar{\ell}(u)$
- $A_3(k, u)$: $gap(u) = \begin{cases} \infty & \text{if } u \preceq s'_{\ell(u)} \\ \overrightarrow{r_{\ell(u)}}(s'_{\ell(u)}) - d_P(s'_{\ell(u)}, u) & \text{otherwise} \end{cases}$
- $A_4(u)$: $\forall i : t \neq s'_i \preceq u \rightarrow u \in V'_i$

The following (somewhat technical) lemma will be used to prove Lemmas 6 and 7. Basically, it states that in the highway query algorithm the search level and the gap to the next applicable neighbourhood border are set cor-rectly.

Lemma 5 *Consider a path $P = \langle s, \ldots, t \rangle$ and an edge (u, v) on P. As-sume that u is* settled *by the highway query algorithm appropriately with some key k. We consider the attempt to relax the edge (u, v). After Line 9 has been executed, the following* <u>I</u>*nvariants apply w.r.t. the variables ℓ and gap:*

- I1: (a) $s_\ell \preceq u \wedge$ (b) $v \preceq s_{\ell+1}$
- I2: $\ell = \bar{\ell}(v)$
- I3: $gap = \begin{cases} \infty & \text{if } v \preceq s'_\ell, \\ \overrightarrow{r_\ell}(s'_\ell) - d_P(s'_\ell, u) & \text{otherwise.} \end{cases}$

Proof. We distinguish between two cases in order to prove I1 and I3.
Case 1: zero iterations of the for-loop in Line 9 take place ($\ell = \ell(u)$). In this case, we have $\ell = \ell(u)$ and $w(u, v) \leq gap'$. Hence, $s_\ell \preceq u$ due to $A_2(k, u)$ and L2 ($\Rightarrow$ I1a). In order to show I1b and I3, we distinguish between three subcases:

- *Subcase 1.1:* $u \prec s'_\ell \Rightarrow v \preceq s'_\ell \preceq s_{\ell+1}$ (U1) ($\Rightarrow$ I1b). Furthermore, because of $\text{gap}(u) = \infty$ ($A_3(u,k)$), we have $\text{gap} = \text{gap}' = r_{\ell(u)}^{\rightarrow}(u) = \infty$ due to U3 and R1 ($\Rightarrow$ I3 since $v \preceq s'_\ell$).

- *Subcase 1.2:* $u = s'_\ell \Rightarrow \text{gap}(u) = \infty$ ($A_3(u,k)$) $\Rightarrow w(u,v) \le \text{gap}' = r_\ell^{\rightarrow}(u)$ (Line 7) $\Rightarrow d_P(s'_\ell, v) \le r_\ell^{\rightarrow}(s'_\ell)$ (since $u = s'_\ell$) $\Rightarrow v \preceq s_{\ell+1}$ (U2b) ($\Rightarrow$ I1b). Furthermore, $\text{gap} = \text{gap}' = r_\ell^{\rightarrow}(u) = r_\ell^{\rightarrow}(s'_\ell) - d_P(s'_\ell, u)$ (since $u = s'_\ell$) implies I3 since $s'_\ell \prec v$.

- *Subcase 1.3:* $u \succ s'_\ell \Rightarrow \text{gap}(u) = r_\ell^{\rightarrow}(s'_\ell) - d_P(s'_\ell, u)$ ($A_3(u,k)$). By Lemma 2, $\ell \le L$ and ($\ell = L \to \text{gap} = \infty$). If $\ell = L$, we have $v \preceq t = s_{L+1} = s_{\ell+1}$ ($\Rightarrow$ I1b) and $\text{gap} = \infty = r_\ell^{\rightarrow}(s'_\ell) - d_P(s'_\ell, u)$ (R2) ($\Rightarrow$ I3 since $s'_\ell \prec v$). Subsequently, we deal with the remaining case $\ell < L$. The facts that $u \preceq t$ and $s'_\ell \prec u$ imply $s'_\ell \ne t$, which yields $s'_\ell \in V'_\ell$ due to U4. Hence, due to R3, $\text{gap}(u) \ne \infty \Rightarrow w(u,v) \le \text{gap}' = \text{gap}(u)$ (Line 7) $\Rightarrow d_P(u,v) \le r_\ell^{\rightarrow}(s'_\ell) - d_P(s'_\ell, u) \Rightarrow d_P(s'_\ell, v) \le r_\ell^{\rightarrow}(s'_\ell)$ $\Rightarrow v \preceq s_{\ell+1}$ (U2b) ($\Rightarrow$ I1b). Furthermore, $\text{gap} = \text{gap}' = \text{gap}(u) = r_\ell^{\rightarrow}(s'_\ell) - d_P(s'_\ell, u)$ implies I3 since $s'_\ell \prec v$.

Case 2: at least one iteration of the for-loop takes place ($\ell > \ell(u)$). We claim that after any iteration of the for-loop, we have $u = s_\ell$. Proof by induction:

Base Case: We consider the first iteration of the for-loop. Line 9 and the fact that an iteration takes place imply $w(u,v) > \text{gap}'$, which means that $\text{gap}' \ne \infty$. We distinguish between two subcases to show that $d_P(s'_{\ell(u)}, v) > r_{\ell(u)}^{\rightarrow}(s'_{\ell(u)})$.

- *Subcase 2.1:* $u \preceq s'_{\ell(u)} \Rightarrow \text{gap}(u) = \infty$ ($A_3(u,k)$) $\Rightarrow w(u,v) > \text{gap}' = r_{\ell(u)}^{\rightarrow}(u)$ (Line 7) $\Rightarrow r_{\ell(u)}^{\rightarrow}(u) \ne \infty$. We have $s_{\ell(u)} \preceq u \preceq s'_{\ell(u)}$ due to L2, $A_2(u,k)$, and the assumption of Subcase 2.1. However, we can exclude that $s_{\ell(u)} \preceq u \prec s'_{\ell(u)}$: this would imply $u \notin V'_{\ell(u)}$ (U3) and, thus, $r_{\ell(u)}^{\rightarrow}(u) = \infty$ (R1). Therefore, $u = s'_{\ell(u)} \Rightarrow d_P(s'_{\ell(u)}, v) > r_{\ell(u)}^{\rightarrow}(s'_{\ell(u)})$

- *Subcase 2.2:* $u \succ s'_{\ell(u)} \Rightarrow s'_{\ell(u)} \ne t \Rightarrow s'_{\ell(u)} \in V'_{\ell(u)}$ (U4). Furthermore, $\text{gap}(u) = r_{\ell(u)}^{\rightarrow}(s'_{\ell(u)}) - d_P(s'_{\ell(u)}, u)$ ($A_3(u,k)$) $\Rightarrow \text{gap}(u) \ne \infty$ (due to R3 since $\ell(u) < L$ (Lemma 2) and $s'_{\ell(u)} \in V'_{\ell(u)}$) $\Rightarrow d_P(u,v) = w(u,v) > \text{gap}' = \text{gap}(u) = r_{\ell(u)}^{\rightarrow}(s'_{\ell(u)}) - d_P(s'_{\ell(u)}, u)$ (Line 7) $\Rightarrow d_P(s'_{\ell(u)}, v) > r_{\ell(u)}^{\rightarrow}(s'_{\ell(u)})$

From $d_P(s'_{\ell(u)}, v) > \overrightarrow{r_{\ell(u)}}(s'_{\ell(u)})$, it follows that $s_{\ell(u)+1} \prec v$ (U2a), which implies $s_{\ell(u)+1} \preceq u$. Hence, $u = s_{\ell(u)+1}$ (since $u \preceq s_{\ell(u)+1}$ due to L3 and $A_2(k, u)$).

Induction Step: Let us now deal with the iteration from level i to level $i+1$ for $i \geq \ell(u)+1$. We have $w(u, v) > \text{gap} = \overrightarrow{r_i}(u)$, which implies $\overrightarrow{r_i}(u) \neq \infty$. Starting with $u = s_i \preceq s'_i \preceq s_{i+1}$ (induction hypothesis, U1), we can conclude that $u = s'_i$ (U3, R1) $\Rightarrow d_P(s'_i, v) > \overrightarrow{r_i}(s'_i) \Rightarrow s_{i+1} \prec v$ (U2a) $\Rightarrow s_{i+1} \preceq u \Rightarrow u = s_{i+1}$ (since $u \preceq s_{i+1}$). This completes our inductive proof.

After the last iteration, we have $u = s_\ell \preceq s'_\ell$ ($\Rightarrow$ I1a). Furthermore, $w(u, v) \leq \overrightarrow{r_\ell}(u)$. If $u \prec s'_\ell$, we obtain $v \preceq s'_\ell \preceq s_{\ell+1}$ ($\Rightarrow$ I1b) and gap $= \overrightarrow{r_\ell}(u) = \infty$ due to U3 and R1 ($\Rightarrow$ I3 since $v \preceq s'_\ell$). Otherwise ($u = s'_\ell$), we get $d_P(s'_\ell, v) \leq \overrightarrow{r_\ell}(s'_\ell)$, which implies $v \preceq s_{\ell+1}$ as well (U2b) ($\Rightarrow$ I1b); furthermore, gap $= \overrightarrow{r_\ell}(u) = \overrightarrow{r_\ell}(s'_\ell) - d_P(s'_\ell, u)$ (since $u = s'_\ell$) implies I3 since $s'_\ell \prec v$. This completes the proof of I1 and I3.

I2 ($\overline{\ell}(v) = \ell$) directly follows from $s_\ell \prec v \preceq s_{\ell+1}$ (I1).   □

Lemma 6 *Consider a highway path $P = \langle s, \ldots, t \rangle$ and an edge (u, v) on P such that u precedes the meeting point p. Assume that u has been appropriately settled. Then, the edge (u, v) is not skipped, but relaxed.*

Proof. We consider the relaxation of the edge (u, v). Due to Lemma 5, the Invariants I1–I3 apply after Line 9 has been executed. Now, we consider Line 10 of the highway query algorithm.

I1 and M2 imply $s_\ell \preceq u \prec p \preceq s_{\lambda+1}$. Hence, $\ell \leq \lambda$. Thus, $u \prec p \preceq t_\lambda \preceq t_\ell$ (M1). By H2, we obtain $\ell(u, v) \geq \ell$. Therefore, the edge (u, v) is not skipped at this point.

Moreover, we prove that the condition in Line 11 is not fulfilled since (u, v) belongs to a highway path. This means that the edge (u, v) is not skipped at this point, either. We have to show that $u \notin V'_\ell \vee v \notin B_\ell$. We have $s_\ell \preceq u$ (I1). If $u \prec s'_\ell$, we get $u \notin V'_\ell$ (U3). Otherwise, we have $s'_\ell \preceq u \prec v \preceq p \preceq t'_\ell$ (M3), which yields $v \notin B_\ell$ (H1).

Therefore, (u, v) is not skipped, but relaxed.   □

Lemma 7 *Consider a shortest path $P = \langle s, \ldots, t \rangle$ and an edge (u, v) on P. Assume that u has been appropriately settled with some key k. Furthermore, assume that the edge (u, v) is not skipped, but relaxed. Then, v can be appropriately reached from u with key k'.*

Proof. We consider the relaxation of the edge (u, v). Due to Lemma 5, the Invariants I1–I3 apply after Line 9 has been executed. Therefore—since (u, v) is not skipped, but relaxed—, the node v can be reached with the key

$$k' = (\delta'(v), \ell'(v), \text{gap}'(v)) := (\delta(u) + w(u, v), \ell, \text{gap} - w(u, v)).$$

Thus, $A_1(k', v)$, $A_2(k', v)$, and $A_3(k', v)$ hold since P is a shortest path and due to $A_1(k, u)$, I2, and I3.

Consider an arbitrary i such that $t \neq s'_i \preceq v$. To prove $A_4(v)$, we have to show that $v \in V'_i$. Due to U4, this is true for $s'_i = v$. Now, we deal with the remaining case $s'_i \preceq u \prec v$. Since $v \preceq s_{\ell+1} \preceq s'_{\ell+1}$ (I1, U1), we have $i \leq \ell$. The case $\ell = 0$ is trivial; hence, we assume $\ell > 0$. Since the edge (u, v) is not skipped, we know that Restriction 1 does not apply. Therefore, we have $\ell(u, v) \geq \ell$, which implies $v \in V_\ell \subseteq V'_{\ell-1}$. For $i < \ell$, we have $V'_{\ell-1} \subseteq V'_i$ and are done. For $i = \ell$, we have $u \in V'_\ell$ due to $A_4(u)$. This implies $v \notin B_\ell$ since Restriction 2 does not apply as well. $v \in V_\ell$ and $v \notin B_\ell$ yield $v \in V'_\ell$. $\quad\square$

Analogous considerations hold for the backward search.

3.5.5   Finding an Optimal Path

Source and target nodes s and t are given such that a shortest path from s to t exists.[3]

Definition 4 *A state z is a triple $(P, u, \overline{u})$, where P is a s-t-path, $u, \overline{u} \in V \cap P$, and $u \preceq \overline{u}$.*

Definition 5 *A state $z = (P, u, \overline{u})$ is* valid *iff all of the following valid State properties are fulfilled:*

- S1: $w(P) = d_0(s, t)$

- S2: $P|_{u \to \overline{u}}$ *is contracted, i.e.,* $P|_{u \to \overline{u}} = \text{ctr}(P|_{u \to \overline{u}})$

- S3: $P|_{s \to u}$ *and* $P|_{\overline{u} \to t}$ *are paths in the forward and backward search tree, respectively.*

---

[3]The special case that there is no path from s to t is treated in Section 3.5.1.

Lemma 8 *Consider a valid state* $z = (P, u, \overline{u})$ *and an arbitrary node* $x, s \preceq x \preceq u$, *on* P. *Then,* x *has been appropriately settled. Analogously for backward search.*

Proof. Base Case: True for s. *Induction Step:* We assume that $y, s \preceq y \prec u$, has been appropriately settled and show that $x = \mathrm{succ}(y)$ is appropriately settled as well. Since (y, x) belongs to the forward search tree (S3), we know that (y, x) is not skipped, but relaxed. The other prerequisites of Lemma 7 are fulfilled as well (due to the induction hypothesis and S1). Thus, we can conclude that x can be appropriately *reached* from y. Since (y, x) belongs to the forward search tree, we know that x is also *settled* from y. $\qquad\square$

Lemma 9 *If* $z = (P, u, \overline{u})$ *is a valid state, then* P *is a highway path.*

Proof. All labels (e.g., s'_ℓ) in this proof refer to P. We show that the highway properties H1 and H2 are fulfilled by induction over the level ℓ.
Base Case: H2(0) trivially holds since $\ell(u, v) \geq 0$ for *any* edge (u, v).
Induction Step (a): H2(ℓ) $\rightarrow$ H1(ℓ). We assume $s'_\ell \prec t'_\ell$. (Otherwise, H1(ℓ) is trivially fulfilled.) This implies $s'_\ell \neq t$. Consider an arbitrary node x on $P|_{s'_\ell \rightarrow t'_\ell}$. We distinguish between three cases.
Case 1: $x \preceq u$. According to Lemma 8, $A_4(x)$ holds. Hence, $x \in V'_\ell$ since $s'_\ell \preceq x$.
Case 2: $u \preceq x \preceq \overline{u}$. We have $y := \max(u, s'_\ell) \in V'_\ell$ (either by Lemma 8: $A_4(u)$ or by U4). Analogously, $\overline{y} := \min(\overline{u}, t'_\ell) \in V'_\ell$. Since $u \preceq y \preceq x \preceq \overline{y} \preceq \overline{u}$ and $P|_{u \rightarrow \overline{u}} = \mathrm{ctr}(P|_{u \rightarrow \overline{u}})$ (S2), we can conclude that $x \notin B_\ell$. Furthermore, we have $x \in V_\ell$ (due to H2(ℓ)). Thus, $x \in V'_\ell$.
Case 3: $\overline{u} \preceq x$. Analogous to Case 1.
Induction Step (b): H1(ℓ) $\wedge$ H2(ℓ) $\rightarrow$ H2($\ell + 1$). Let $\overline{P}$ denote $\exp(P|_{s'_\ell \rightarrow t'_\ell}, \ell)$ and consider an arbitrary edge (x, y) on $\overline{P}$. If (x, y) is part of an expanded shortcut, we have $\ell(x, y) \geq \ell + 1$ and $x, y \in V_{\ell+1} \subseteq V'_\ell$. Otherwise, (x, y) belongs to $P|_{s'_\ell \rightarrow t'_\ell}$, which is a subpath of $P|_{s_\ell \rightarrow t_\ell}$, which implies $x, y \in V'_\ell$ and $\ell(x, y) \geq \ell$ by H1(ℓ) and H2(ℓ). Thus, in any case, $\ell(x, y) \geq \ell$, $x, y \in V'_\ell$, and (x, y) is not a shortcut of some level $> \ell$. Hence, $\overline{P}$ is a path in G'_ℓ. Now, consider an arbitrary edge (u, v), $s_{\ell+1} \preceq u \prec v \preceq t_{\ell+1}$, on P. If (u, v) is a shortcut of some level $> \ell$, we directly have $\ell(u, v) \geq \ell + 1$. Otherwise, (u, v) is on $\overline{P}$ as well. Since $s_{\ell+1} \prec v$, we

have $d_P(s'_\ell, v) > r_\ell^{\rightarrow}(s'_\ell)$ (U2b). Moreover, S1 implies that $\overline{P}$ is a shortest path in G'_ℓ and, in particular, $d_{\overline{P}}(s'_\ell, v) = w(\overline{P}|_{s'_\ell \rightarrow v}) = d_\ell(s'_\ell, v)$. Using the fact that $d_{\overline{P}}(s'_\ell, v) = d_P(s'_\ell, v)$, we obtain $d_\ell(s'_\ell, v) > r_\ell^{\rightarrow}(s'_\ell)$ and, thus, $v \notin \mathcal{N}_\ell^{\rightarrow}(s'_\ell)$.

Analogously, we have $u \notin \mathcal{N}_\ell^{\leftarrow}(t'_\ell)$. Hence, the definition of the high-way network $G_{\ell+1}$ implies $(u, v) \in E_{\ell+1}$. Thus, $\ell(u, v) \geq \ell + 1$. $\qquad \square$

Definition 6 *A* valid state *is either a* final state *(if $u = \overline{u}$) or a* non-final state *(otherwise).*

We pick any shortest s-t-path P. The state $(\text{ctr}(P), s, t)$ is the *initial* state. Since forward and backward search run completely independently of each other, any serialisation of both search processes will yield exactly the same result. Therefore, in our proof, we are free to pick—w.l.o.g.—any order of forward and backward steps. We assume that at first one forward and one backward iteration is performed, which implies that s and t are settled. At this point, the highway query algorithm is in the initial state. It is easy to see that the initial state is a valid state. Due to the following lemma, it is sufficient to prove that a final state is eventually reached.

Lemma 10 *Getting to a final state is equivalent to finding a shortest s-t-path.*

Proof. $u = \overline{u}$ means that forward and backward search meet. Due to Lemma 8, we can conclude that both u and $\overline{u}$ are settled with the optimal distance (A_1), i.e., $\overrightarrow{\delta}(u) = d_0(s, u)$ and $\overleftarrow{\delta}(\overline{u}) = d_0(\overline{u}, t)$. Since $u = \overline{u}$ lies on a shortest path (due to S1), we have $d(s, t) = d_0(s, u) + d_0(\overline{u}, t)$. Line 6 implies $d' \leq \overrightarrow{\delta}(u) + \overleftarrow{\delta}(\overline{u}) = d(s, t)$. In fact, this means that the algorithm returns $d' = d(s, t)$ since this is already optimal. $\qquad \square$

Definition 7 *For a valid state $z = (P, u, \overline{u})$, the* forward direction *is said to be* blocked *if $p \preceq u$. Analogously, the* backward direction *is blocked if $\overline{u} \preceq p$.*

Lemma 11 *For a non-final state $z = (P, u, \overline{u})$, at most one direction is blocked.*

Proof. Since z is a non–final state, we have $u \prec \overline{u}$, which implies $u \prec p$ or $p \prec \overline{u}$. □

Definition 8 *The* rank $\rho(z)$ *of a state* $z = (P, u, \overline{u})$ *is* $|\{x \in P \mid u \preceq x \preceq \overline{u}\}|$.

Lemma 12 *From any non-final state* $z = (P, u, \overline{u})$, *another valid state* z^+ *is reached at some point such that* $\rho(z^+) < \rho(z)$.

Proof. We pick any non–blocked direction—due to Lemma 11, we know that there is at least one such direction. Subsequently, we assume that the forward direction was picked; the backward direction can be dealt with anal–ogously.

We have $u \prec p$ and observe that all prerequisites of Lemma 6 are ful–filled due to Lemmas 9 and 8. Hence, we can conclude that the edge $(u, v := \mathrm{succ}(u))$ is not skipped, but relaxed. Thus, since P is a shortest path (S1), v can be reached with the optimal distance due to Lemma 7 (A_1). The fact that the algorithm terminates (Lemma 3) implies that the queue $\overrightarrow{Q}$ gets empty at some point, i.e., every element has been deleted from $\overrightarrow{Q}$. In particular, we can conclude that v is deleted at some point. Since v has been reached with the optimal distance, it will also be settled with the optimal distance (due to the specification of the decreaseKey operation, which guarantees that ten–tative distances are never increased). Let P' denote the path from s to v in the forward search tree. We set $z^+ := (P^+ := P' \circ P|_{v \to t}, v, \overline{u})$. We have $w(P^+) = w(P') + w(P|_{v \to t}) = d_0(s, v) + d_0(v, t) = d_0(s, t)$ ($\Rightarrow$ S1). S2 is fulfilled since $P^+|_{v \to \overline{u}}$ is a subpath of $P|_{u \to \overline{u}}$. S3 holds due to the construction of P^+. Hence, z^+ is valid. Furthermore, $\rho(z^+) = \rho(z) - 1$. □

Theorem 5 *The highway query algorithm finds a shortest s-t-path.*

Proof. From Lemma 12 and the fact that the codomain of the rank function is finite, it follows that eventually a final state is reached, which is equivalent to finding a shortest s–t–path due to Lemma 10. □

3.5.6 Distance Table Optimisation

To prove the correctness of the distance table optimisation, we introduce the following new lemma and adapt a few definitions and proofs from Section 3.5.5 to the new situation.

Lemma 13 *Consider a valid state* $z = (P, u, \overline{u})$ *with* $u \prec s'_L$. *When* u's *edges are relaxed, neither the condition in Line 7a nor the condition in Line 11a is fulfilled.*

Proof. Due to Lemma 8, u has been appropriately settled with some key k. We distinguish between two cases.

Case 1: $u \prec s_L$. From $s_{\ell(u)} = s_{\overline{\ell}(u)} \preceq u \prec s_L$ ($A_2(k, u)$, L2), it follows that $\ell(u) < L$ (U1). Hence, the condition in Line 7a is not fulfilled. Furthermore, we have $s_\ell \preceq u \prec s_L$ after Line 9 has been executed (Lemma 5: I1). Thus, $\ell < L$, which implies that the condition in Line 11a is not fulfilled as well.

Case 2: $s_L \preceq u \prec s'_L$. First, we show that the condition in Line 7a is not fulfilled. We assume $\ell(u) = L$. (Otherwise, the condition cannot be fulfilled.) Due to $A_3(k, u)$, we have $\mathrm{gap}(u) = \infty$. Hence, $\mathrm{gap}' = r_{\overrightarrow{\ell(u)}}(u) = r_{\overrightarrow{L}}(u) = \infty_1$ by R1 since $u \notin V'_L$ (U3). Now, we prove that the condition in Line 11a is not fulfilled. We assume $\ell = L \wedge \ell > \ell(u)$. (Otherwise, the condition cannot be fulfilled.) Due to Line 9, we get $\mathrm{gap} = r_{\overrightarrow{\ell}}(u) = r_{\overrightarrow{L}}(u) = \infty_1$ (as above). □

Definition 6'. *A valid state is either a* final *state (if* $u = \overline{u}$ *or* $s'_L \preceq u \wedge \overline{u} \preceq t'_L$) *or a* non-final *state (otherwise).*

Lemma 10. *Getting to a final state is equivalent to finding a shortest s-t-path.*

Proof. In the proof of this lemma in Section 3.5.5, we have already dealt with the case $u = \overline{u}$. Now, consider the new case $u \prec \overline{u} \wedge s'_L \preceq u \wedge \overline{u} \preceq t'_L$. We show that s'_L is added to the set $\overrightarrow{I}$. Since $s'_L \preceq u$, s'_L has been appropriately settled with some key k (due to Lemma 8). We consider the attempt to relax the edge $(s'_L, v := \mathrm{succ}(s'_L))$ and distinguish between two cases.

Case 1: $s_L = s'_L$. $\ell = \overline{\ell}(v)$ (I2), $s_L = s'_L \prec v$, and $\overline{\ell}(v) \leq L$ (L1) imply $\ell = \overline{\ell}(v) = L$. Furthermore, $A_2(k, s'_L)$ and the assumption of Case 1 yield $\ell(s'_L) = \overline{\ell}(s'_L) < L = \ell$. In addition, gap $= \infty_2 \neq \infty_1$ by I3 (since $s'_\ell \prec v$), the fact that $s'_L \in V'_L$ (U4), and R2. Hence, the condition in Line 11a is fulfilled so that s'_L is added to $\overrightarrow{T}$.

Case 2: $s_L \prec s'_L$. By $A_2(k, s'_L)$, $A_3(k, s'_L)$, the assumption of Case 2, and $\overline{\ell}(s'_L) \leq L$ (L1), we get $\ell(s'_L) = \overline{\ell}(s'_L) = L$ and gap$(s'_L) = \infty$. Thus, gap$' = r_L^{\rightarrow}(s'_L) = \infty_2 \neq \infty_1$ (R2). Hence, the condition in Line 7a is fulfilled so that s'_L is added to $\overrightarrow{T}$.

Analogously, we can prove that t'_L is added to the set $\overleftarrow{T}$. Since P is a highway path (due to Lemma 9), the subpath $P|_{s'_L \to t'_L}$ is a path in G'_L and, thus, $d_0(s'_L, t'_L) = d_L(s'_L, t'_L)$. Hence, $w(P) = d_0(s, s'_L) + d_L(s'_L, t'_L) + d_0(t'_L, t)$ is the length of a shortest s–t–path and, since the algorithm finds a path with a length $\leq \overrightarrow{\delta}(s'_L) + d_L(s'_L, t'_L) + \overleftarrow{\delta}(t'_L)$ and since $\overrightarrow{\delta}(s'_L) = d_0(s, s'_L)$ and $\overleftarrow{\delta}(t'_L) = d_0(t'_L, t)$ (due to Lemma 8: A_1), we can conclude that a shortest s–t–path is found.   □

Definition 7'. *For a valid state* $z = (P, u, \overline{u})$, *the forward direction is said to be* blocked *if* $p \preceq u$ *or* $s'_L \preceq u$. *Analogously, the backward direction is* blocked *if* $\overline{u} \preceq p$ *or* $\overline{u} \preceq t'_L$.

Lemma 11. *For a non-final state* $z = (P, u, \overline{u})$, *at most one direction is blocked.*

Proof. Since z is a non-final state, we have $u \prec \overline{u}$ and $(u \prec s'_L \vee t'_L \prec \overline{u})$. To obtain a contradiction, let us assume that both directions are blocked, i.e., $(p \preceq u$ or $s'_L \preceq u)$ and $(\overline{u} \preceq p$ or $\overline{u} \preceq t'_L)$. Consider the case $p \preceq u$ and $\overline{u} \preceq t'_L$. Hence, $p \preceq u \prec \overline{u} \preceq t'_L$. Due to M3, we can conclude that $s'_L \preceq p \preceq u$. Since $s'_L \preceq u$ and $\overline{u} \preceq t'_L$, we have a contradiction. The remaining three cases are analogous or straightforward.   □

Lemma 12. *From any non-final state* $z = (P, u, \overline{u})$, *another valid state* z^+ *is reached at some point such that* $\rho(z^+) < \rho(z)$.

Proof. The proof of this lemma in Section 3.5.5 still works since the added two lines (7a and 11a) have no effect due to Definition 7' and Lemma 13. □

3.6   Combination with Goal-Directed Search

The highway query algorithm is *not* goal-directed. In fact, the forward search 'knows' nothing about the target and the backward search 'knows' nothing about the source, so that both search processes work completely independently and spread into all directions.

In order to obtain further speedups, a combination with a goal-directed approach seems promising. Subsequently, we study a combination with A^* search using landmarks.

3.6.1 A^* Search Using Landmarks

In this section we recapitulate the well-known technique of A^* search [35] and in particular the ALT algorithm [32], which is a specialisation of A^* search using so-called *landmarks*. As a new result, we present how the selection of landmarks can be accelerated using highway hierarchies.

A^* **Search.**   The search space of Dijkstra's algorithm can be visualised as a circle around the source. The idea of A^* search is to push the search towards the target. By adding a potential $\pi : V \to \mathbb{R}$ to the priority of each node, the order in which nodes are removed from the priority queue is altered. A 'good' potential lowers the priority of nodes that lie on a shortest path to the target. It is easy to see that A^* is equivalent to Dijkstra's algorithm on a graph with *reduced costs*, formally $w_\pi(u, v) = w(u, v) - \pi(u) + \pi(v)$. Since Dijkstra's algorithm works only on nonnegative edge costs, not all potentials are allowed. We call a potential π *feasible* if $w_\pi(u, v) \geq 0$ for all $(u, v) \in E$. The distance from each node v of G to the target t is the distance from v to t in the graph with reduced edge costs minus the potential of t plus the potential of v. So, if the potential $\pi(t)$ of the target t is zero, $\pi(v)$ provides a *lower bound* for the distance from v to the target t.

Bidirectional A^*.   At a glance, combining A^* and bidirectional search seems easy. Simply use a feasible potential π_f for the forward and a feasible potential π_r for the backward (or '*reverse*') search. However, this does not work due to the fact that both searches might work on different reduced costs so that the shortest path might not have been found when both searches meet.

This can only be guaranteed if π_f and π_r are *consistent* meaning $w_{\pi_f}(u, v)$ in G is equal to $w_{\pi_r}(v, u)$ in the reverse graph. We use the variant of an average potential function [40] defined as $p_f(v) = (\pi_f(v) - \pi_r(v))/2$ for the forward and $p_r(v) = (\pi_r(v) - \pi_f(v))/2 = -p_f(v)$ for the backward search. Note that these potentials are feasible and consistent, but provide worse lower bounds than the original ones.

ALT. There exist several techniques [81, 98] how to obtain feasible potentials using the layout of a graph. The ALT algorithm uses a small number of nodes—so-called *landmarks*—and the triangle inequality to compute feasible potentials. Given a set $S \subseteq V$ of landmarks and distances $d(\mathcal{L}, v), d(v, \mathcal{L})$ for all nodes $v \in V$ and landmarks $\mathcal{L} \in S$, the following triangle inequations hold:

$$d(u, v) + d(v, \mathcal{L}) \geq d(u, \mathcal{L}) \quad \text{and} \quad d(\mathcal{L}, u) + d(u, v) \geq d(\mathcal{L}, v).$$

Therefore, $\underline{d}(u, v) := \max_{\mathcal{L} \in S} \max\{d(u, \mathcal{L}) - d(v, \mathcal{L}), d(\mathcal{L}, v) - d(\mathcal{L}, u)\}$ provides a lower bound for the distance $d(u, v)$. The quality of the lower bounds highly depends on the quality of the selected landmarks.

Considering *all* landmarks for the computation of a lower bound is time–consuming. Instead, for each s–t query only two landmarks—one 'before' the source and one 'behind' the target—are initially used. At certain check–points it is decided whether to add an additional landmark to the set of active landmarks.

Landmark-Selection. A crucial point in the success of ALT is the quality of landmarks. Since finding good landmarks is hard, several heuristics [29, 32] exist. One technique that provides particular good landmarks is *maxCover*. Unfortunately, its application is rather expensive: Calculating maxCover landmarks on our Western European road network takes about 75 minutes, while constructing the whole highway hierarchy can be done in about 15 minutes. A promising approach is to use the highway hierarchy to reduce the number of possible landmarks: The level–1 core of the European network has six times fewer nodes than the original network and its construction takes only about three minutes. Using the core as possible positions for landmarks, the computation time for calculating landmarks can be decreased.[4] Using only the nodes of higher level cores reduces the time for selecting landmarks even more. Figure 3.15 shows an example of 16 landmarks, generated on the level-1 core of the European network.

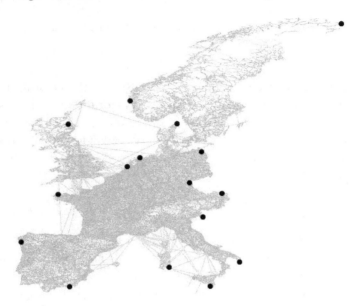

Figure 3.15: 16 core-1 landmarks on the Western European road network.

3.6.2 Combining Highway Hierarchies and A^* Search

Previously (see Section 3.4.2), we strictly separated the search phase to the topmost core from the access to the distance table: first, the sets of entrance points $\overrightarrow{I}$ and $\overleftarrow{I}$ into the core of the topmost level were determined, and afterwards the table look-ups were performed. Now we interweave both phases: whenever a forward core entrance point u is discovered, it is added to $\overrightarrow{I}$ and we immediately consider all pairs $(u, v), v \in \overleftarrow{I}$, in order to check whether the tentative shortest path length d' can be improved. (An analogous procedure applies to the discovery of a backward core entrance point.) This new approach is advantageous since we can use the tentative shortest path length d' as an upper bound on the actual shortest path length.

---

[4]This applies to all known heuristics, not only to maxCover.

In [69, 70], the highway query algorithm used a strategy that compares the minimum elements of both priority queues and prefers the smaller one in order to sequentialise forward and backward search. If we want to obtain good upper bounds very fast, this might not be the best choice. For example, if the source node belongs to a densely populated area and the target to a sparsely populated area, the distances from the source and target to the entrance points into the core of the topmost level will be very different. Therefore, we now choose a strategy that balances $|\overrightarrow{I}|$ and $|\overleftarrow{I}|$, preferring the direction that has encountered less entrance points. In case of equality (in particular, in the beginning when $|\overrightarrow{I}| = |\overleftarrow{I}| = 0$), we use a simple alternating strategy.

We enhance the highway query algorithm with goal–directed capabilities—obtaining an algorithm that we call HH* *search*—by replacing edge weights by *reduced costs* using potential functions π_f and π_r for forward and backward search. By this means, the search is directed towards the respective target, i.e., we are likely to find some s–t path very soon. However, just using the reduced costs only changes the *order* in which the nodes are settled, it does not reduce the search space. The ideal way to benefit from the early encounter of the forward and backward search would be to abort the search as soon as an s–t path has been found. And, as a matter of fact, in case of the ALT algorithm [29]—even in combination with reach–based routing [26]—it can be shown that an immediate abort is possible without losing correctness if consistent potential functions are used. In contrast, this does not apply to the highway query algorithm since even in the non–goal–directed variant of the algorithm, we cannot abort when both search scopes have met (see Section 3.4.2).

Fortunately, there is another aspect of goal–directed search that can be exploited, namely *pruning*: finding any s–t path also means finding an upper bound d' on the length of the shortest s–t path. Comparing the lower bounds with the upper bound can be used to prune nodes: if the key of a settled node u is greater than the upper bound, we do not have to relax u's edges. Note that, using reduced costs, the key of u is the distance from the corresponding source to u plus the lower bound on the distance from u to the corresponding target.

Since we do not abort when both search scopes have met and because we have the distance table, a very simple implementation of the ALT algo–

rithm is possible. First, we do not have to use consistent potential functions. Instead, we directly use the lower bound to the target as potential for the for–ward search and, analogously, the lower bound from the source as potential for the backward search. These potential functions make the search pro–cesses approach their respective target faster than using consistent potential functions so that we get good upper bounds very early. In addition, the node pruning gets very effective: if one node is pruned, we can conclude that all nodes left in the same priority queue will be pruned as well since we use the same lower bound for pruning and for the potential that is part of the key in the priority queue. Hence, in this case, we can immediately stop the search in the corresponding direction.

Second, it is sufficient to select at the beginning of the query for each search direction only one landmark that yields the best lower bound. Since the search space is limited to a relatively small local area around source and target (due to the distance table optimisation), we do not have to pick more landmarks, in particular, we do not have to add additional landmarks in the course of the query, which would require flushing and rebuilding the priority queues. Thus, adding A^* search to the highway query algorithm (including the distance table optimisation) causes only little overhead per node.

However, there is a considerable drawback. While the goal–directed search (which gives good upper bounds) works very well, the pruning is not very successful when we want to compute *fastest* paths, i.e., when we use a travel time metric, because then the lower bounds are usually too weak. Figure 3.16 gives an example for this observation, which occurs quite fre–quently in practice. The first part of the shortest path from s to t corresponds to the first part of the shortest path from s to the landmark u. Thus, the re–duced costs of these edges are zero so that the forward search starts with traversing this common subpath. The backward search behaves in a similar way. Hence, we obtain a perfect upper bound very early (a). Still, the lower bound on $d(s,t)$ is quite bad: we have $d(s,u) - d(t,u) \le d(s,t)$. Since staying on the motorway and going directly from s to u is much faster than leaving the motorway, driving through the countryside to t and continuing to u, the distance $d(s,t)$ is clearly underestimated. The same applies to lower bounds on $d(v,t)$ for nodes v close to s. Hence, pruning the forward search does not work properly so that the search space still spreads into all directions before the process terminates (b). In contrast, the node s lies on the shortest path (in the reverse graph) from t to the landmark that is used by the backward search. (Since this landmark is very far away to the south, it has not been included in the figure.) Therefore, the lower bound is perfect so that the backward search stops immediately. However, this is a fortunate case that occurs rather rarely.

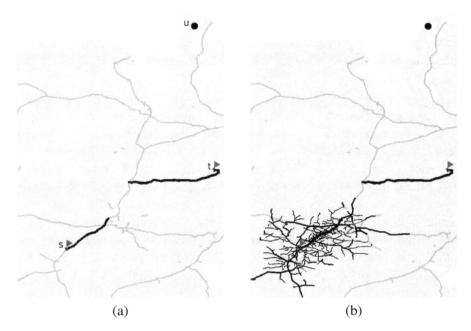

(a)   (b)

Figure 3.16: Two snapshots of the search space of an HH* search using a travel time metric. The landmark u of the forward search from s to t is explicitly marked. The landmark used by the backward search is somewhere below s and not included in the chosen clipping area. The search space is black, parts of the shortest path are represented by thick lines. In addition, motorways are highlighted.

Approximate Queries. We pointed out above that in most cases we *find* a (near) shortest path very quickly, but it takes much longer until we *know* that the shortest path has been found. We can adapt to this situation by defining an abort condition that leads to an approximate query algorithm: when a node u is removed from the forward priority queue and we have

$(1 + \varepsilon) \cdot (d(s, u) + \underline{d}(u, t)) > d'$ (where $\varepsilon \geq 0$ is a given parameter), then the search is not continued in the forward direction. In this case, we may miss some s-t-paths whose length is $\geq d(s, u) + \underline{d}(u, t)$ since the key of any remaining element v in the priority queue is $\geq d(s, u) + \underline{d}(u, t)$ and it is a lower bound on the length of the shortest path from s via v to t. Thus, if the shortest path is among these paths, we have $d(s, t) \geq d(s, u) + \underline{d}(u, t) > d'/(1+\varepsilon)$, i.e., we have the guarantee that the best path that we have already found (whose length corresponds to the upper bound d') is at most $(1 + \varepsilon)$ times as long as the shortest path. An analogous stopping rule applies to the backward search.

Better Upper Bounds. We can use the distance table to get good upper bounds even earlier. So far, the distance table has only been applied to entrance points into the core V_L' of the topmost level. However, in many cases we encounter nodes that belong to V_L' earlier during the search process. Even the source and the target node could belong to the core of the topmost level. Still, we have to be careful since the distance table only contains the shortest path lengths within the topmost core and a path between two nodes in V_L' might be longer if it is restricted to the core of the topmost level instead of using all edges of the original graph. This is the reason why we have not used such a premature jump to the highest level before. But now, in order to just determine upper bounds, we could use these additional table look-ups. The effect is limited though because finding good upper bounds works very well anyway—the lower bounds are the crucial part. Therefore, the exact algorithm does without the additional look-ups. The approximate algorithm applies this technique to the nodes that remain in the priority queues after the search has been terminated since this might improve the result.[5] For example, we would get an improvement if the goal-directed search led us to the wrong motorway entrance ramp, but the right entrance ramp has at least been inserted into the priority queue.

Reducing Space Consumption. We can save preprocessing time and memory space if we compute and store only the distances between the landmarks and the nodes in the core of some fixed level k. Obviously, this has

---

[5]In a preliminary experiment, the total error observed in a random sample was reduced from 0.096% to 0.053%.

the drawback that we cannot begin with the goal-directed search immediately since we might start with nodes that do not belong to the level-k core
so that the distances to and from the landmarks are not known. Therefore,
we introduce an additional *initial query phase*, which works as a normal
highway query and is stopped when all entrance points into the core of level
k have been encountered. Then, we can determine the distances from s to
all landmarks since the distances from s via the level-k core entrance points
to the landmarks are known. Analogously, the distances from the landmarks
to t can be computed. The same process is repeated for interchanged source
and target nodes—i.e., we search forward from t and backward from s—in
order to determine the distances from t to the landmarks and from the landmarks to s. Note that this second subphase can be skipped when the first
subphase has encountered only bidirected edges.

The priority queues of the *main query phase* are filled with the entrance
points that have been found during (the first subphase of) the initial query
phase. We use the distances from the source or target node plus the lower
bound to the target or source as keys for these initial elements. Since we
never leave the level-k core during the main query phase, all required distances to and from the landmarks are known and the goal-directed search
works as usual. The final result of the algorithm is the shortest path that has
been found during the initial or the main query phase.

3.7 Concluding Remarks

Review. Highway hierarchies are the first route planning technique that
was able handle the road network of a whole continent, achieving speedups
of more than a factor 1 000 compared to Dijkstra's algorithm. They offer a
good compromise between preprocessing time, memory consumption, and
query time. In particular w.r.t. preprocessing time, they are superior to practically any other method that achieves significant speedups. The facts that
they can handle all types of queries efficiently, that we can give per-instance
worst-case guarantees, and that a few tuning parameters can be used to
obtain different trade-offs between memory consumption and query times
make highway hierarchies applicable in a wide range of applications. The
most significant drawback, however, is that—in contrast to highway-node
routing (Chapter 4)—(so far) highway hierarchies cannot handle dynamic

scenarios accurately.[6]

In addition to being useful by themselves, highway hierarchies inspired various other speedup techniques or even constitute a starting point for them. In particular, this is true for the methods presented in Chapters 4–6. More-over, Goldberg et al. adopted the shortcut concept in order to improve both preprocessing and query times of reach–based routing (Section 1.2.3).

References. This chapter is based on [75, 69, 70, 17, 68]. In order to be self–contained, we gave a complete account on highway hierarchies that also covers parts that have already been included in the Master's thesis [75] and thus, are not an official part of this thesis due to formal reasons. The main advancements of the current version compared to the version from [75] are listed in Section 1.3.2. The combination of highway hierarchies with landmark–based A^* search [17] was a joint work with Daniel Delling (among others), who particularly attended to the landmark–specific parts, while the combination itself can be seen as a (non–exclusive) part of this thesis.

---

[6]There is a *heuristic* approach based on highway hierarchies by Nannicini et al. [65].

4

Highway-Node Routing

4.1 Central Ideas

Let us assume that we have identified 'important' nodes V'—which we call *highway nodes*—in a given road network and furthermore, let us assume for a moment that we want to compute shortest paths only between these highway nodes. One extreme (and trivial) solution would be to just do the routing in the original graph, the other extreme would be to precompute a $|V'| \times |V'|$ distance table. A third possibility is to construct an *overlay graph*, i.e., a graph that consists of the node set V' and of an edge set such that the distance in the overlay graph between any node pair $(u, v) \in V' \times V'$ agrees with the corresponding distance in the original network. Note that a distance table can be represented as an overlay graph if we introduce an edge with the appropriate weight between each node pair. Usually, however, we are interested in an overlay graph with a minimal edge set. Figure 4.1 gives an example.

Computing routes between highway nodes using an overlay graph is usually much faster than doing the routing in the original graph. But such an overlay graph is even more useful: we can specify a simple bidirectional query algorithm that works for any node pair $(s, t) \in V \times V$. It is somewhat similar to the query procedure of highway hierarchies and consists of two phases: First, we search forwards from the source s and backwards from the target t until the respective search tree is *covered* by nodes from V', i.e., each branch of the tree contains at least one node from V'. Second, we continue the search in the (hopefully considerably smaller) overlay graph from the nodes in V' that have been settled during the first phase. An extension of this general idea to multiple levels suggests itself.

We could think of various ways to determine the highway nodes, which are obviously a crucial ingredient for an implementation of this new route planning technique. One good possibility is to use the classification that is provided by our highway hierarchies approach: nodes in high levels of a highway hierarchy can be considered as more important than nodes in lower levels. The construction of an overlay graph (for a given node set V') can be done by performing a search from each node $u \in V'$ until all branches of the search tree are covered. We pick on each branch the node $v \in V'$ closest to the root u and add an edge (u, v) with the distance that corresponds to

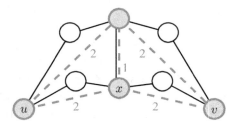

Figure 4.1: A minimal overlay graph (shaded nodes, dashed edges, ex–plicitly given edge weights) of the depicted graph (solid edges, unit edge weights). Note that a direct edge (u, v) is not required since a shortest path from u to v is already represented by the edges (u, x) and (x, v).

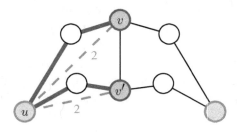

Figure 4.2: A local search as part of an overlay graph construction process. The search is started from u, thick edges belong to the search tree, shaded nodes belong to V'. The search can be stopped when v and v' have been settled. Two edges (u, v) and (u, v') (dashed) are added to the overlay graph.

the length of the path from u to v in the search tree. For an example, refer to Figure 4.2. It turns out that certain tricks are required to get an approach that works efficiently in real–world road networks, which contain some nasty elements like long–distance ferry connections.

4.2 Covering Paths

The concept of *covering paths* plays a central role for highway–node routing. Before using it extensively in Section 4.3, we introduce it here in a general way.

4.2.1 Canonical Shortest Paths

For a given graph $G = (V, E)$, $\mathcal{U}(G)$ is a set of *canonical shortest paths* if it contains for each connected pair $(s, t) \in V \times V$ exactly one unique shortest path from s to t such that $P = \langle s, \ldots, s', \ldots, t', \ldots, t \rangle \in \mathcal{U}(G)$ implies that $P|_{s' \to t'} \in \mathcal{U}(G)$.

It is easy to see that Dijkstra's algorithm always finds canonical shortest paths if we have a total order on the nodes and, in case of ambiguities, prefer the parent node with the smaller rank.

4.2.2 Basics

Covering-Paths Set. We consider a graph $G = (V, E)$, a node subset $V' \subseteq V$, a node $s \in V$, and a set $C \subseteq \{\langle s, \ldots, u \rangle \mid u \in V'\}$ of paths in G.

Definition 9 *The set C is a* covering–paths set *of s w.r.t. V' if for any node $t \in V'$ that can be reached from s, there is a node $u \in V'$ on some shortest s-t-path P such that $P|_{s \to u} \in C$, i.e.,*

$$P = \langle s, \ldots, \overbrace{u}^{\in V'}, \ldots, \overbrace{t}^{\in V'} \rangle.$$
$$\underbrace{}_{\in C}$$

Definition 10 *A covering-paths set C is a* canonical covering–paths set *if for any node $t \in V'$ that can be reached from s, there is a node $u \in V'$ on the <u>canonical</u> shortest s-t-path P such that $P|_{s \to u} \in C$.*

Note that $\{P = \langle s, \ldots, u \rangle \mid P \in \mathcal{U}(G) \wedge u \in V'\}$ is a trivial canonical covering–paths set.

The crucial subroutine of all algorithms in the subsequent sections takes a graph G, a node set V', and a root s and determines a set of canonical covering paths. In the process, there are two conflicting objectives: the computation should be as fast as possible and the resulting set should be as small as possible. In the following, we present a generic algorithm and four concrete instantiations that allow different trade-offs between these objectives.

Local Search.   We consider Dijkstra's algorithm that has been modified as described above so that it determines canonical shortest paths. During a Dijkstra search from s, we say that a settled node u is *covered* by a node set V' if there is at least one node $v \in V'$ on the path from the root s to u. A queued node is *covered* by V' if its tentative parent is covered by V'. The current search tree B is *covered* by V' if all currently queued nodes are covered by V'.

A *local Dijkstra search* is a Dijkstra search with an additional pruning rule that allows for not continuing the search from certain nodes that are covered or that have been settled on a suboptimal path. "Not continuing" the search from a node u means that u's edges are not relaxed when u is settled. Subsequently, we will introduce four different concrete pruning rules, but at first, in the following lemma, we deal with the general case.

Lemma 14 *Consider a node set V' and a local Dijkstra search from a node s that yields a search tree B. We define C to consist of all paths $\langle s, \ldots, v \rangle$ in B with an endpoint $v \in V'$ that has no parent in B that is covered by V'. Then, C is a canonical covering-paths set of s w.r.t. V'.*

Proof. Consider any node $t \in V'$ that can be reached from s and the canonical shortest s–t–path P. If P contains no covered node, the search cannot have been pruned at any node on P since P is a shortest path. Hence, P belongs to B and, thus, to C as well since $t \in V'$ does not have a covered parent in B. Otherwise (if there is a covered node on P), consider the first covered node v on P that does not have a covered parent. Such a node always exists since the root s has no parent at all. We can conclude that all nodes $u \prec v$ on P are not covered and, consequently, that the search has not been pruned at any node $u \prec v$. Hence, $P|_{s \to v}$ belongs to B. Moreover, $v \in V'$. Hence, $P|_{s \to v} \in C$. $\square$

4.2.3 Conservative Approach

The *conservative* variant (Figure 4.3 (a)) works in the obvious way: the search from s is stopped (i.e., all remaining nodes in the queue are pruned) as soon as the current search tree B is covered. This yields a canonical covering-paths set close to the optimum.[1] However, if B contains one path that is not covered for a long time, B can get very big even though all other branches might have been covered very early. Therefore, it takes a long time until the local search terminates. In our application, this is a critical issue in particular due to long-distance ferry connections.

4.2.4 Aggressive Approach

As an overreaction to the above observation, we might want to define an *aggressive* variant that prunes the search at every covered node, i.e., some branches can be terminated early, while only the non-covered paths are followed further on. Unfortunately, this provokes two problems. First, the covering-paths set gets unnecessarily big.[2] Second, the tree B can get even bigger since the search might continue *around* the nodes where we pruned the search.[3] In our example (Figure 4.3 (b)), the search is pruned at u so that v is reached using a much longer path that leads around u. As a consequence, the path to w is superfluously marked as a covering path.

4.2.5 Stall-in-Advance Technique

If we decide not to prune the search immediately, but to go on 'for a while' in order to *stall* other branches, we obtain a compromise between the conservative and the aggressive variant, which we call *stall-in-advance*. One heuristic we use prunes the search at node z when the path explored from s to z contains a nodes of V' for some tuning parameter a. Note that for

---

[1]In order to guarantee that the minimal set is found, we would have to slightly change the tie-breaking rule that decides the case that a node can be reached on different shortest paths: if there is a choice, we have to prefer the already covered parent. This variant has been introduced and proven correct in [37, 38].

[2]In Section 4.3.3, we will explain how to reduce such a covering-paths set rather efficiently in order to obtain a minimal set.

[3]Note that the separator-based approach virtually uses the aggressive variant. This is reasonable since the search can never 'escape' the component where it started.

$a := 1$, the stall-in-advance variant corresponds to the aggressive variant. In our example (Figure 4.3 (c)), we use $a := 2$. Therefore, the search is pruned not until w is settled. This stalls the edge (s, v) and, in contrast to (b), the node v is covered. Still, the search is pruned too early so that the edge (s, x) is used to settle x.

4.2.6   Stall-on-Demand Technique

In the stall-in-advance variant, relaxing an edge leaving a covered node is based on the 'hope' that this might stall another branch. However, our heuristic is not perfect, i.e., some edges are relaxed in vain, while other edges which would have been able to stall other branches, are not relaxed. Since we are not able to make the perfect decision in advance, we introduce a fourth variant, namely *stall-on-demand*. It is an extension of the aggressive variant, i.e., the search is pruned immediately at all covered nodes. We introduce a new concept, called *stalling process*. It can be started from a node u that has been pruned earlier. The goal is to identify nodes that have been reached on a suboptimal path. We can prove that this is the fact for a node v if there is a path from s via u to v that is shorter than the best path to v

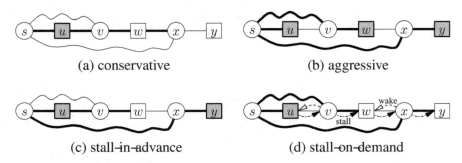

(a) conservative   (b) aggressive

(c) stall-in-advance   (d) stall-on-demand

Figure 4.3: Simple example for the computation of covering paths. We assume that all edges have weight 1 except for the edges (s, v) and (s, x), which have weight 10. In each case, the search process is started from s. The set V' consists of all nodes that are represented by a square. Thick edges belong to the search tree B. Endpoints of the computed covering paths are highlighted in grey. Note that the minimal covering-paths set contains only the path to u.

found so far. In order to find such *witness paths*, we perform a search from u considering reached nodes. This search is continued only from nodes whose suboptimality can be proven. Such nodes are marked as *stalled*. The main search is pruned not only at covered nodes (as mentioned above), but also at stalled nodes. A stalling process from u can be invoked by an adjacent node v—we say that v *wakes* u—if there is an edge (u, v) that is relaxed from v.[4] In our example (Figure 4.3 (d)), the search is pruned at u. When v is settled, we assume that the edge (v, w) is relaxed first. Then, the edge (v, u) wakes the node u. A stalling process is started from u. The nodes v and w are marked as stalled. When w is settled, its outgoing edges are not relaxed. Similarly, the edge (x, w) wakes the stalled node w and another stalling process is performed.

Algorithmic Details. We implement the stalling process as a breadth–first search (BFS) instead of a shortest–path search since a BFS causes less over–head and yields almost the same stalling effect (as preliminary experiments indicate). We store the length of the corresponding witness path at each stalled node. If such a node is woken up later, we can start the stalling process based on this witness path (instead of the path that was found orig–inally, which is longer and, thus, less qualified to stall further nodes). Note that if we mark a queued node as stalled, it can happen that it is reached later on (before it is settled) by the main search on a shorter path. In this case, we have to remove the stalled–marker. As an optimisation, we do not add paths whose endpoint has been stalled to the covering–paths set. It is obvi–ous that omitting such suboptimal paths does not invalidate the correctness of Lemma 14.

4.3 Static Highway-Node Routing

4.3.1 Multi-Level Overlay Graph

Overlay Graph. An overlay graph of a given graph consists of a subset of the nodes and an edge set that has the property that shortest path distances are preserved. More formally, for a given graph G_ℓ and a node set $V_{\ell+1}$, the

---

[4]We only wake up u if there is a backward edge from v to u because if there was only a forward edge (v, u), it could not be part of a witness path from s via u to v.

graph $G_{\ell+1} = (V_{\ell+1}, E_{\ell+1})$ is an *overlay graph* of G_ℓ if for all $(u, v) \in V_{\ell+1} \times V_{\ell+1}$, we have $d_{\ell+1}(u, v) = d_\ell(u, v)$, where $d_\ell(u, v) := d_{G_\ell}(u, v)$ denotes the distance from u to v in G_ℓ. One way to define the edge set $E_{\ell+1}$ with the desired property is to use the covering–paths concept as it is shown in the following lemma. Note that this is similar to the corresponding definition in [37, 38].

Lemma 15 *Consider a graph G_ℓ and a node set $V_{\ell+1}$. For any node $u \in V_{\ell+1}$, let $C(u)$ denote a canonical covering-paths set of u w.r.t. $V_{\ell+1} \setminus \{u\}$. Then, $G_{\ell+1} := (V_{\ell+1}, \{(u, v) \mid u \in V_{\ell+1} \wedge \langle u, \ldots, v \rangle \in C(u)\})$ with $w(u, v) := w(\langle u, \ldots, v \rangle)$ is an overlay graph of G_ℓ.*

Proof. We consider an arbitrary node pair $(u, v) \in V_{\ell+1} \times V_{\ell+1}$. Since all edges in $E_{\ell+1}$ represent paths in G_ℓ, we have $d_{\ell+1}(u, v) \geq d_\ell(u, v)$. Thus, if $d_\ell(u, v) = \infty$, we also have $d_{\ell+1}(u, v) = \infty$. Subsequently, we assume $d_\ell(u, v) \neq \infty$. We do an inductive proof over the number i of nodes from $V_{\ell+1}$ on the canonical shortest u–v–path.
Base Case: $i = 1$, i.e., $u = v$. Trivial.
Induction Step: $1, \ldots, i \rightarrow i + 1$. Consider a node pair $(u, v) \in V_{\ell+1} \times V_{\ell+1}$ whose canonical shortest u–v–path P has $i + 1$ nodes from $V_{\ell+1}$. The definition of the canonical covering–paths set $C(u)$ implies that there is a node $x \in V_{\ell+1} \setminus \{u\}$ on P such that $P|_{u \rightarrow x} \in C(u)$. From $P \in \mathcal{U}(G_\ell)$, it follows that $P|_{x \rightarrow v} \in \mathcal{U}(G_\ell)$. Moreover, $P|_{x \rightarrow v}$ contains at most i nodes from $V_{\ell+1}$. Due to the induction hypothesis, we have $d_{\ell+1}(x, v) = d_\ell(x, v)$.

The definition of $G_{\ell+1}$ implies that there is an edge $(u, x) \in E_{\ell+1}$ with $w(u, x) = w(P|_{u \rightarrow x}) = d_\ell(u, x)$. Summing up, we have $d_{\ell+1}(u, v) \leq w(u, x) + d_{\ell+1}(x, v) = d_\ell(u, x) + d_\ell(x, v) = d_\ell(u, v)$, which implies $d_{\ell+1}(u, v) = d_\ell(u, v)$. $\square$

Multi-Level Overlay Graph. The overlay graph definition can be applied iteratively to define a multi–level hierarchy. For given *highway-node sets* $V =: V_0 \supseteq V_1 \supseteq \ldots \supseteq V_L$, we define the *multi-level overlay graph* $\mathcal{G} = (G_0, G_1, \ldots, G_L)$ in the following way: $G_0 := G$ and for each $\ell \geq 0$, $G_{\ell+1}$ is the overlay graph of G_ℓ. The *level* $\ell(u)$ of a node $u \in V$ is $\max\{\ell \mid u \in V_\ell\}$. Analogously, the *level* $\ell(u, v)$ of an edge $(u, v) \in \bigcup_{i=0}^{L} E_i$ is $\max\{\ell \mid (u, v) \in E_\ell\}$. Again, these definitions are similar to those in [37, 38].

4.3.2 Node Selection

We can choose any highway–node sets to get a correct procedure. However, this choice has a big impact on preprocessing and query performance.

Let us consider a Dijkstra search from some node in a road network. We observe that some branches are very important—they extend through the whole road network—, while other branches are stalled by the more important branches at some point. For instance, there might be all types of roads (motorways, national roads, rural roads) that leave a certain region around the source node, but usually the branches that leave the region via rural roads end at some point since all further nodes are reached on a faster path using motorways or national roads. We want to exploit this observation: not all nodes that separate different regions are selected as highway nodes, but only the nodes on the important branches. Note that this is a crucial distinction from the separator–based multi–level method.[5] In order to classify the nodes by importance, we employ our highway hierarchies approach: we use the set of level-ℓ core nodes of the highway hierarchy of G as highway–node set V_ℓ.

4.3.3 Construction

The multi–level overlay graph is constructed in a bottom–up fashion. In order to construct level $\ell > 0$, we determine for each node $u \in V_\ell$ a canonical covering–paths set $C(u)$ in $G_{\ell-1}$ w.r.t. $V_\ell \setminus \{u\}$ using one of the methods from Section 4.2 and apply Lemma 15.

Edge Reduction. Optionally, we can apply the following *reduction step* to eliminate edges from E_ℓ that are superfluous: for each node $u \in V_\ell$, we perform a search in G_ℓ (instead of $G_{\ell-1}$) until all adjacent nodes have been settled. If there is a tie during the search, we prefer the path that contains more nodes. Then, we can remove any edge (u, v) whose target v has been settled via a path that consists of more than one edge since a (better) alternative that does not require the edge (u, v) has been found.

---

[5]More details on differences from the separator–based approach can be found in Section 4.3.5.

Lemma 16 *Applying the edge reduction step to an overlay graph G_ℓ of $G_{\ell-1}$ yields a minimal overlay graph G'_ℓ of $G_{\ell-1}$.*

Proof.[6] Obviously, removing only edges (u, v) from G_ℓ that can be replaced by using a different path P from u to v with $w(P) \le w(u, v)$ does not invalidate the overlay graph property. We still have to show minimality. Assume that there is an overlay graph G''_ℓ with $|E''_\ell| < |E'_\ell|$. Hence, there is some edge (u, v) in $E'_\ell \setminus E''_\ell$. Since G'_ℓ and G''_ℓ are overlay graphs of the same graph $G_{\ell-1}$, distances in G'_ℓ are equal to distances in G''_ℓ. Hence, there is some shortest u–v-path P'' in G''_ℓ with $w(P'') \le w(u, v)$. From $(u, v) \notin E''_\ell$, it follows that P'' has at least one interior node x. Furthermore, the shortest path P' in G'_ℓ from u via x to v has the same length as P''. Thus, $w(P') = w(P'') \le w(u, v)$. This implies that (u, v) is removed by the edge reduction step, i.e., $(u, v) \notin E'_\ell$, which is a contradiction. □

4.3.4 Query

Level-Synchronised Variant. The query algorithm is a symmetric Dijkstra–like bidirectional procedure, which works in a bottom–up fashion. To get a first intuition, we can think of the following algorithm that is syn–chronised by search level. We give a description only for the forward search. First, we search in level 0, i.e., we determine a covering–paths set of the source node s in G_0 w.r.t. V_1 using one of the methods from Section 4.2. Then, we search level 1, i.e., for each endpoint u of a covering path, we determine the covering–paths set $C(u)$ of u in G_1 w.r.t. V_2. After that, we search level 2, starting from all endpoints of paths in $\bigcup C(u)$. And so on.

Forward and backward search are interleaved. We keep track of a ten–tative shortest–path length (which is initially set to ∞) resulting from nodes that have been settled in both search directions. We abort the forward (back–ward) search when all keys in the forward (backward) priority queue are greater than the tentative shortest–path length.

Asynchronous, Aggressive Variant. The level–synchronised variant de–scribed above would be rather inefficient since it does not pay attention to the fact that the search reaches the level borders in an irregular way: while

---

[6]This proof has been inspired by the proof of Theorem 2.2 in [38].

at one side of the search frontier the covering paths might be found very
early, this might not be the case at a different side. We should allow the
search process to proceed in higher levels even if the search in a lower level
has not been completed yet; in other words, we prefer an asynchronous variant. It is convenient to describe such an asynchronous algorithm based on
the aggressive approach (Section 4.2.4) in the following way: We define the
forward search graph

$$\overrightarrow{\mathcal{G}} := (V, \{(u, v) \mid (u, v) \in E_{\ell(u)}\})$$

and, analogously, the *backward search graph*

$$\overleftarrow{\mathcal{G}} := (V, \{(u, v) \mid (v, u) \in E_{\ell(u)}\}).$$

We perform two *normal* Dijkstra searches in $\overrightarrow{\mathcal{G}}$ and in $\overleftarrow{\mathcal{G}}$. As in the level–
synchronised variant, forward and backward search are interleaved, we keep
track of a tentative shortest–path length and abort the forward/backward
search process not until all keys in the respective priority queue are greater
than the tentative shortest–path length. Note that we are *not* allowed to abort
the entire query as soon as both search scopes meet for the first time. This
is due to similar reasons as in the case of highway hierarchies (cp. Section 3.4.2).

   Although it might not be obvious at first glance, this algorithm is based
on the aggressive approach because whenever a node in a higher level ℓ
is settled, we immediately switch to that higher level (by considering only
edges in level ℓ); in other words, we immediately prune[7] the search in the
lower level. Figure 4.4 illustrates the query algorithm and gives an intuition
for its correctness. Consider a shortest s–t path P in the original graph
(level 0) and the first and last nodes s_1 and t_1 on P that belong to V_1. Due
to the definition of the overlay graph G_1 (level 1), we have $d_1(s_1, t_1) = d_0(s_1, t_1)$. The same argument can be applied iteratively; in our example,
we also have $d_2(s_2, t_2) = d_1(s_2, t_2)$. Now, consider the path P' from s to s_1
in the original graph, continuing in G_1 to s_2, continuing in G_2 to t_2, in G_1 to
t_1, and finally in the original graph to t. It is easy to see that $w(P') = w(P)$
and that the first part of P' up to t_2 belongs to the forward search graph $\overrightarrow{\mathcal{G}}$,

---

[7]Note that pruning is done *implicitly* due to the definition of the forward and backward
search graphs.

while the last part starting with s_2 belongs to the backward search graph $\overleftarrow{\mathcal{G}}$. Since s_2 and t_2 belong to both parts, they can be settled from both sides so that the shortest path P' is found.

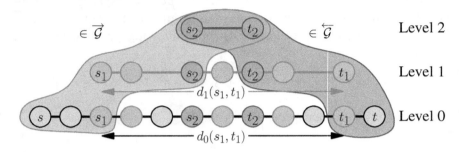

Figure 4.4: Illustration of the query algorithm.

Theorem 6 *The asynchronous, aggressive query algorithm is correct.*

Proof. The query algorithm terminates since Dijkstra's algorithm always terminates. If there is no shortest path from source to target in G, the algorithm will correctly return ∞ since no node is settled from both search directions: this is due to the fact that only paths in the multi-level overlay graph are considered and these paths correspond to paths in the original graph. Now, consider a node pair $(s_0, t_0) \in V \times V$ with $d_0(s_0, t_0) \neq \infty$ and some shortest s_0–t_0–path P_0 in G_0. Note that, trivially, $\ell(x, y) \geq 0$ for any edge (x, y), $s_0 \preceq x \prec y \preceq t_0$. For each level ℓ from 0 to $L - 1$ consider the following steps: If $P_\ell \cap V_{\ell+1} = \emptyset$, break. Otherwise, let $s_{\ell+1}$ and $t_{\ell+1}$ denote the first and the last node from $V_{\ell+1}$ on P_ℓ, respectively. ($s_{\ell+1}$ and $t_{\ell+1}$ can be equal.) Since $G_{\ell+1}$ is an overlay graph of G_ℓ, we have $d_{\ell+1}(s_{\ell+1}, t_{\ell+1}) = d_\ell(s_{\ell+1}, t_{\ell+1})$. Pick a shortest $s_{\ell+1}$–$t_{\ell+1}$–path in $G_{\ell+1}$ and replace $P_\ell|_{s_{\ell+1} \to t_{\ell+1}}$ with it; we obtain $P_{\ell+1}$. Note that $w(P_{\ell+1}) = w(P_\ell)$. Since $P_{\ell+1}|_{s_{\ell+1} \to t_{\ell+1}}$ is a path in $G_{\ell+1}$, we have $\ell(x, y) \geq \ell + 1$ for all edges (x, y) on $P_{\ell+1}$ between $s_{\ell+1}$ and $t_{\ell+1}$.

After the last iteration, where we defined P_i, we set $P := P_i$ and the *meeting point* p to t_i. Now, it is easy to see that $w(P) = d_0(s_0, t_0)$, $P|_{s \to p}$ belongs to $\overrightarrow{\mathcal{G}}$, and the reverse path of $P|_{p \to t}$ belongs to $\overleftarrow{\mathcal{G}}$. The forward search is not aborted before p is settled because up to this point, the tentative shortest path distance is at least $w(P)$ and the minimum key in the priority queue is at most $w(P|_{s \to p}) \leq w(P)$. The same argument applies to the backward search. Hence, p is settled from both search directions, which implies that the algorithm returns the correct result.   □

Note that we may either discard all edges that do not belong to the forward or backward search graph and perform the search on the remain–ing graph (which is exactly what we described above) or we can keep all edges and make sure that only edges that belong to the search graph are re–laxed by introducing an explicit level check. The former variant is simpler, more space–efficient, and faster since we can do without the additional level checks. The latter variant, however, is more flexible since keeping all edges and level data allows modifications to the multi–level overlay graph as they are needed in dynamic scenarios (Section 4.4).

We can also obtain a *unidirectional* variant of the query algorithm using techniques originally introduced to handle dynamic scenarios. For details, see Section 4.4.2.

Stall-on-Demand. The integration of the stall–on–demand technique (Sec–tion 4.2.6) is straightforward. Note that if we have kept only the forward and backward search graphs, then a small problem arises: Let us consider the forward search. When a node v wakes a node u so that u can stall v, then the edge (u, v) does not belong to the forward search graph. Thus, if we just started the stalling process from u, it would fail. Fortunately, there is a simple workaround: since the reverse edge (v, u) belongs to the backward search graph and is used to wake u, the required data is available and we only have to make sure that it is used appropriately.

Obviously, pruning the search at nodes that have been demonstrably settled via a suboptimal path cannot invalidate the correctness proven in Theorem 6 since a continuation of the search from these nodes can never contribute to a shortest path.

Outputting Complete Path Descriptions. In order to output a complete description of the computed shortest path, we have to unpack the edges of the overlay graphs in order to obtain the represented subpaths in the original graph. This can be done using the same techniques as for highway hierar–chies (see Section 3.4.3).

4.3.5 Analogies and Differences To Related Techniques

In this section, we want to compare the highway–node routing approach with two related techniques, pointing out some analogies and differences. The goal is to convey some *intuition* of the relation between the discussed techniques, without making accurate and provable propositions.

Separator-Based Multi-Level Method. In contrast to highway–node routing, in the separator–based multi–level approach *all* nodes that separate different regions are selected, which leads to a comparatively high average node degree. This has a negative impact on the performance. Let us consider the 'old' multi–level method with the new selection strategy, i.e., only 'important' nodes are selected. Then, the graph is typically not decomposed into many small components so that the following performance problem arises in the query algorithm. From the highway/separator nodes, only edges of the overlay graph are relaxed. As a consequence, the unimportant branches are not stalled by the important branches. Thus, since the separator nodes on the unimportant branches have not been selected, the search might extend through large parts of the road network.

To sum up, there are two major steps to get from the separator–based method to our approach: first, select only 'important' nodes and second, at highway/separator nodes, do not switch immediately to the next level, but keep relaxing low–level edges 'for a while' until you can be sure that slow branches have been stalled (stall–in–advance, Section 4.2.5) or employ the stall–on–demand technique (Section 4.2.6).

Highway Hierarchies. We use the preprocessing of the highway hierarchies in order to select the highway nodes for our new approach. However, this is not the sole connection between both methods. In fact, we can interpret highway–node routing as some kind of special case of the highway hierarchy approach. In this paragraph, we denote the highway–node sets by $S_0, S_1, \ldots, S_L$ instead of $V_0, V_1, \ldots, V_L$ to avoid notational conflicts with the highway hierarchies. For given highway–node sets, consider the following highway hierarchy: To construct the highway network $G_{\ell+1}$ of a graph G'_ℓ (for $0 \le \ell < L$), we set the neighbourhood radii of any node $u \in V'_\ell$ to zero. This virtually means that $G_{\ell+1} = G'_\ell$. To specify the core G'_ℓ of a highway network G_ℓ (for $1 \le \ell \le L$), we set B_ℓ to $V_\ell \setminus S_\ell$, which implies $V'_\ell = S_\ell$. Note that due to its definition, the core G'_ℓ is an overlay graph of G_ℓ (though not a minimal one). A query in this highway hierarchy would basically settle the same nodes as a highway–node query without using the stall–on–demand technique, which is an important add–on of highway–node routing.

In a sense, highway–node routing is a logical advancement of highway hierarchies: We started with the highway hierarchies concept as presented in Section 3.2, i.e., we iteratively first construct a highway network and then contract it. In our experiments (see Section 7.4.1), however, we observed that we get particularly good results when starting with a contraction step, followed by alternating construction and contraction steps. In order to start with a contraction step without changing the implementation, we set the neighbourhood radii of all nodes to zero so that the first construction step had no effect and we virtually started with the contraction of the original graph. In case of the highway–node routing approach, this 'trick' is now applied to all levels.

Note that our highway–node routing implementation reuses large parts of the highway hierarchies program code. However, we do not just reduce highway–node routing to a special case of highway hierarchies since, for example, dealing with trivial neighbourhood radii would cause unnecessary overhead.

4.4 Dynamic Highway-Node Routing

Subsequently, we deal with two different dynamic scenarios. First, we want to switch to a different cost function, which means that potentially all edge weights change. For example, a cost function can take into account differ–ent weightings of travel time, distance, scenic value, and fuel consumption. With respect to travel time, we can think of different profiles of average speeds for each road category. In addition, for certain vehicle types there might be restrictions on some roads (e.g., bridges and tunnels).

Second, we want to cope with unexpected incidents, like traffic jams, which influence the expected travel time of a certain road or several roads in some area. That means, a single or a few edge weights change. Here, we can distinguish between a *server scenario* and a *mobile scenario*: In the former, a server has to react to incoming events by updating its data structures so that *any* point–to–point query can be answered correctly; in the latter, a mobile device has to react to incoming events by (re)computing a *single* point–to–point query taking into account the new situation. In the server scenario, it pays to invest some time to perform the update operation since a lot of queries depend on it. In the mobile scenario, we do not want to waste time for updating parts of the graph that are irrelevant to the current query.

4.4.1 Changing the Entire Cost Function

The more time–consuming part of the preprocessing is the determination of the highway–node sets. We observe that, when we switch to a different 'reasonable' cost function, properties of the road network (like the inher–ent hierarchy) are possibly weakened, but not completely destroyed or even inverted. For instance, both a truck and a sports car—despite going differ–ent speeds—drive faster on a motorway than on an urban street. Thus, we can still expect a good query performance when *keeping* the highway–node sets and *completely recomputing* only the overlay graphs. In order to do so, we do not need any additional data structures. We can directly use the static approach from Section 4.3 omitting the first preprocessing step (the determination of the highway–node sets).

4.4.2 Changing a Few Edge Weights

Server Scenario. Similar to an exchange of the cost function, when a sin–gle or a few edge weights change, we keep the highway–node sets and up–date only the overlay graphs. In this case, however, we do not have to repeat the complete construction from scratch, but it is sufficient to perform the construction step only from nodes that might be affected by the change. Certainly, a node v whose search tree of the initial construction does not contain any node u of a modified edge (u, x) is *not* affected: if we repeated the construction step from v, we would get exactly the same search tree and, consequently, the same result.

During the first construction (and all subsequent update operations), we manage sets A_u^ℓ of nodes whose level–ℓ preprocessing might be affected when an outgoing edge of u changes: when a level–ℓ construction step from some node v is performed, we add v to A_u^ℓ for each node u in the search tree whose edges are relaxed.[8] Note that these sets can be stored explicitly (as we do it in our current implementation) or we could store a superset, e.g., by some kind of *geometric container* (a disk, for instance). Figure 4.5 contains the pseudo–code of the update algorithm.

> *input*: set of edges E^m with modified weight
>
> define set of modified nodes: $V_0^m := \{u \mid (u, v) \in E^m\}$;
> **foreach** level $\ell, 1 \leq \ell \leq L$, **do**
> $\qquad V_\ell^m := \emptyset; R_\ell := \bigcup_{u \in V_{\ell-1}^m} A_u^\ell$;
> $\qquad$ **foreach** node $v \in R_\ell$ **do**
> $\qquad\qquad$ repeat construction step from v;
> $\qquad\qquad$ if something changes, put v to V_ℓ^m;

Figure 4.5: The update algorithm that deals with a set of edge weight changes.

Mobile Scenario: Single Pass.   In the mobile scenario, we only determine the sets R_ℓ of potentially unreliable nodes by using a fast variant of the update algorithm (Figure 4.5), where from the last two lines only the "put v to V_ℓ^m" is kept. (Note that in particular the construction step is *not* repeated.) Then, for each node $u \in V$, we define the *reliable level* $r(u) := \min\{i - 1 \mid u \in R_i\}$ with $\min \emptyset := \infty$. In order to get correct results without updating the data structures, the query algorithm has to be modified. First, we do not relax any edge (u, v) that has been created during the construction of some level $> r(u)$. Second, if the search at some node u has already reached a level $\ell > r(u)$, then the search at this node is *downgraded* to level $r(u)$. In other words, if we arrive at some node from which we would have to repeat the construction step, we do not use potentially unreliable edges, but continue the search in a sufficiently low level to ensure that the correct path can be found. We call this the *prudent query algorithm*.

---

[8] When the stall–in–advance technique is used, some edges are only relaxed to potentially stall other branches. Upon completion of the construction step, we can identify edges that have been relaxed in vain, i.e., that were not able to stall other branches. Those edges (x, y) had no actual influence on the construction and, thus, we need not add v to A_x^ℓ.

Note that the update procedure, which is also used to determine the sets of potentially unreliable nodes, is performed in the forward direction. Its results cannot be directly applied to the backward direction of the query. It is simple to adjust the first modification to this situation (by considering $r(v)$ instead of $r(u)$ when dealing with an edge (u, v)). Adjusting the second modification would be more difficult, but fortunately we can completely omit it for the backward direction. As a consequence, the search process becomes asymmetric. While the forward search is continued in lower levels whenever it is necessary, the backward search is never downgraded. If 'in doubt', the backward search stops and waits for the forward search to finish the job.

More formally, we redefine the forward and backward search graphs in the following way:

$$\overrightarrow{\mathcal{G}} := (V, \{(u, v) \mid (u, v) \in E_{\min(\ell(u), r(u))}\}),$$

$$\overleftarrow{\mathcal{G}} := (V, \{(u, v) \mid (v, u) \in E_{\ell(u)} \wedge r(v) \geq \ell(u)\}).$$

An example is given in Figure 4.6. We assume that the weight of the edge (x, t_2) has been changed, i.e., $E^m = \{(x, t_2)\}$ and $V_0^m = \{x\}$. Furthermore, we have $A_x^1 = \{x\}$ and $A_x^2 = \{s_2, t_2\}$—these sets have been determined during the construction of the overlay graphs. Applying the pared–down version of the algorithm in Figure 4.5 then yields $R_1 = \{x\}$, $V_1^m = \{x\}$, and $R_2 = \{s_2, t_2\}$. Consequently, we have $r(x) = 0, r(s_2) = r(t_2) = 1$; the reliable level of all other nodes is ∞.

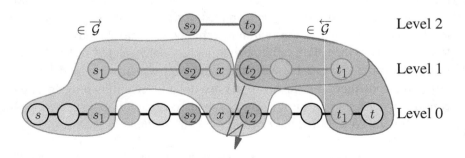

Figure 4.6: Illustration of the query algorithm in case of a single edge weight change (mobile scenario, single pass).

Up to s_2, the forward search can proceed as before (cp. Figure 4.4). Then, since $r(s_2) = 1$, the edge (s_2, t_2) in level 2 cannot be relaxed, but the search stays in level 1. At x the search is even downgraded to level 0 so that the edge (x, t_2) is relaxed in the original graph, where it has its correct up–to–date weight. After that, the forward search can rise again and continue in level 1. The backward search, however, stops already at t_2. According to the (re)definition of $\overleftarrow{\mathcal{G}}$, neither (t_2, s_2) nor (t_2, x) belongs to the backward search graph $\overleftarrow{\mathcal{G}}$. Still, the shortest path is found since both search scopes overlap.

Mobile Scenario: Iterative Variant. Alternatively, we can use an itera–tive variant of the above approach, provided that we allow only edge weight *increases*, which is a very reasonable assumption in particular when we consider traffic jams. Note that the previous methods do *not* rely on this assumption.

Initially, we mark all modified edges so that we can easily decide whether a given edge has been modified. Then, we set the set E^m of mod–ified edges to $\emptyset$. Now, we apply the same algorithm as in the single–pass variant, i.e., determine the reliable levels and perform the prudent query that takes the reliable levels into account. It is easy to see that all reliable levels are ∞ so that a normal query is executed, which returns a path length d'. Of course, this procedure does not necessarily yield correct results. Therefore, we have to examine the computed path: we determine the represented path in the original graph and its length $\hat{d}$ in the original graph, which contains the up–to–date edge weights. If $d' = \hat{d}$, we know that the computed path does not contain any modified edge. Therefore, since we allowed only edge weight increases, it must be a shortest path. Otherwise, we have $d' < \hat{d}$, i.e., the path has become longer due to an edge weight update. We add all marked edges on the path to the set E^m and repeat the computation.

In other words, during the first iteration, we ignore all traffic jams, dur–ing the second iteration, we consider only the traffic jams on the path found in the first iteration, in the third iteration, we consider the traffic jams on the paths found in the first two iterations, and so on. In the worst case, we need as many iterations as modified edges exist. Still, in real–world scenarios, the iterative variant shows promise since usually only a small fraction of all current traffic jams affects the own route.

Unidirectional Query. Interestingly, we can exploit the concepts intro–duced above to obtain a *unidirectional* query algorithm (which can be used in both static and dynamic scenarios). We just have to add the target t to V_0^m (irrespective of the fact whether some edge (t, x) has been modified or not). Then, we compute the reliable levels (exactly as above) and apply the prudent query algorithm. Here, it is sufficient to run the query only in the forward direction. When the forward search approaches the target, it will automatically go down to lower levels and finally reach the target since the edges that would jump over the target have been declared as non–reliable. An example is given in Figure 4.7. We assume that there is no modified edge so that $E^m = \emptyset$ and $V_0^m = \{t\}$. Note that we have $A_t^1 = \{t_1\}$ and $A_{t_1}^2 = \{t_2\}$. Hence, $R_1 = \{t_1\}$, $R_2 = \{t_2\}$, and thus, $r(t_1) = 0$ and $r(t_2) = 1$. Consequently, at t_2, the forward search is downgraded from level 2 to level 1, and at t_1 from level 1 to level 0 so that t is settled by the forward search.

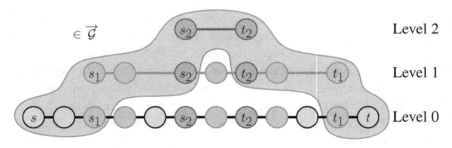

Figure 4.7: Illustration of the unidirectional query algorithm.

Note that in order to apply the unidirectional query algorithm, we still have to know the target node in advance since it is required for the appro–priate determination of the reliable levels.

4.5 Concluding Remarks

Review. Our experiments (Section 7.6) will confirm that the static variant of highway–node routing has outstandingly low memory requirements, fast preprocessing times, and good query performance. Moreover, it is concep–tually very simple. The query algorithm just corresponds to bidirectional

Dijkstra enhanced by one straightforward subroutine (the stall-on-demand technique) that can be implemented in a few lines of code.

More important and innovative, however, is the fact that the approach can be extended to work in dynamic scenarios: we can react to events like traffic jams and we can efficiently switch between different cost functions.

Highway-node routing can also be used to instantiate our generic many-to-many algorithm (Chapter 5) and our generic transit-node routing (Chapter 6), yielding very efficient implementations in both cases.

Future Work. There is room to improve the performance both at the implementation and at the algorithmic level. In particular, better ways to select the highway-node sets might be found: preliminary experiments using a new approach indicate that by this means query times and memory consumption can be further reduced. The preprocessing can be effectively parallelised since the required local searches can be performed independently of each other. The memory consumption of the dynamic variant can be considerably reduced by using a more space-efficient representation of the affected node sets.

Although we have already considered a mobile scenario in case that a few edge weights change, we do not have an implementation optimised for a mobile device yet. However, since the search spaces of highway-node routing are very small and since the hierarchical and geographical structure allows a favourable arrangement of nodes into memory blocks, we expect that an efficient realisation is possible.

At the end of Section 4.4.2, we presented a unidirectional variant of the query algorithm. Unfortunately, this algorithm cannot be applied directly to time-dependent scenarios where the arrival time (and thus, the exact target node in a time-expanded representation of the network) is not known in advance. It is an open question whether we can achieve good performance when computing the reliable levels w.r.t. *several* target nodes (e.g., for each possible arrival time at the given target location).

References. This chapter is based on [76], but also contains several improvements that have not been published yet.

5

Many-to-Many Shortest Paths

5.1 Central Ideas

In contrast to Chapters 3, 4, and 6, here, we do not deal with the point–to–point variant of the shortest–path problem, but with the many–to–many variant, i.e., we are given two node sets S and T and we want to compute the shortest–path distances between *all* node pairs $(s, t) \in S \times T$. In order to solve this problem, we could just employ, for example, highway–node routing once for each node pair, meaning that we execute $|S| \times |T|$ queries. Highway–node routing is a bidirectional technique with the additional prop–erty that forward and backward search proceed completely independently of each other. That means that in the above example, we would execute $|S|$ times the same backward search and $|T|$ times the same forward search.[1] This observation is the starting point for an efficient many–to–many algo–rithm. We decide to perform each backward search only once, storing the resulting search spaces in an appropriate way so that each forward search (which is executed only once as well) can access the deposited information on the backward search spaces.

More precisely, we manage a $|S| \times |T|$ distance table D, where we ini–tially set all entries to infinity. For each node u that a backward search from t encounters, we leave an entry $(t, \overleftarrow{\delta}(u))$ in some bucket that is associated with u. (As introduced before, $\overleftarrow{\delta}(u)$ denotes the computed distance from

---

[1] provided that we omit the abort–on–success criterion, i.e., that we do not stop the searches early after forward and backward search have met

u to t.) During a forward search from a node $s \in S$, we scan the bucket of each visited node u: For each entry $(t, \overleftarrow{\delta}(u))$, we add up the just computed distance $\overrightarrow{\delta}(u)$ from s to u and the stored distance $\overleftarrow{\delta}(u)$ from u to t. The sum represents the length of a path from s via u to t. If it is less than the value $D[s, t]$ stored in the table, we improve the table entry. After having considered all encountered intermediate nodes u, we can be sure that the table contains the correct distances to all targets $t \in T$.

This algorithm can be further improved by introducing some asymmetry: It is not necessary that backward *and* forward searches explore the entire topmost level of the multi-level overlay graph. We can safely prune the backward searches at nodes that belong to the topmost level, i.e., from these nodes we do not continue the search within the topmost level. The correctness is not invalidated since the forward searches will find the shortest paths to the nodes where the backward searches were pruned. By this means, the backward searches get cheaper and, more importantly, there are significantly less bucket entries that the forward searches have to scan. In fact, for large distance tables, the time spent for bucket scanning can become dominating so that it is reasonable to strengthen the asymmetry, i.e., to choose a lower level as topmost level: this increases the forward search spaces, but reduces the number of bucket entries.

Highway-node routing is not the only candidate for constituting the basis of our many-to-many algorithm. Interestingly, a whole class of shortest-path algorithms, which we label as bidirectional, Dijkstra-like, and target-oblivious, comes into question. In the next section, we formally introduce this class—which the generic version of our many-to-many algorithm (presented in Section 5.3) relies on. More details on the concrete instantiation based on highway-node routing (or on the closely related highway hierarchies) then can be found in Section 5.4.

5.2 Bidirectional Target-Oblivious Search

Definition 11 *A* bidirectional Dijkstra-like shortest-path algorithm *is an algorithm that determines for given source and target nodes s and t a forward search space $\overrightarrow{\sigma}(s, t) \subseteq V \times \mathbb{R}_0^+$ and a backward search space $\overleftarrow{\sigma}(s, t) \subseteq V \times \mathbb{R}_0^+$ such that*

$$d(s, t) = \min\{x + y \mid (u, x) \in \overrightarrow{\sigma}(s, t) \wedge (u, y) \in \overleftarrow{\sigma}(s, t)\}. \quad (5.1)$$

Note that $\min \emptyset := \infty$. In the general case, both the forward and the back-ward search space depend on both s and t. For instance, consider the bidirectional version of Dijkstra's algorithm, which is, of course, the most obvious example for a bidirectional Dijkstra-like shortest-path algorithm. If $s = t$, we have $\overrightarrow{\sigma}(s,t) = \{s\}$, which is usually not the case if we choose a different target. Thus, $\overrightarrow{\sigma}(s,t)$ depends on t.

Definition 12 *A bidirectional Dijkstra-like shortest-path algorithm is target-oblivious iff*

$$\forall(s,t_1,t_2) \in V^3 : \overrightarrow{\sigma}(s,t_1) = \overrightarrow{\sigma}(s,t_2) \quad and$$
$$\forall(s_1,s_2,t) \in V^3 : \overleftarrow{\sigma}(s_1,t) = \overleftarrow{\sigma}(s_2,t). \tag{5.2}$$

In other words, the forward search of a target-oblivious algorithm 'knows' nothing about the target, i.e., it proceeds irrespective of the chosen target. An analogous statement applies to the backward search. Consequently, when dealing with target-oblivious algorithms, we can just write $\overrightarrow{\sigma}(s)$ and $\overleftarrow{\sigma}(t)$ instead of $\overrightarrow{\sigma}(s,t)$ and $\overleftarrow{\sigma}(s,t)$.

Bidirectional Dijkstra-like algorithms usually employ some kind of abort-on-success criterion, taking into account the length d' of the best shortest path found so far.[2] An algorithm that uses such an abort-on-success criterion cannot be target-oblivious since the forward (backward) search space size depends on the occurrence of success, which, in turn, depends on the target (source). This fact has already been illustrated in a previous example, where we considered bidirectional Dijkstra and the cases $s = t$ and $s \neq t$.

Moreover, any goal-directed approach cannot be target-oblivious for obvious reasons. Note that this is not a complete exclusion list. For instance, the *bidirectional bound* variant of reach-based routing [26] is not target-oblivious, either, since the pruning rule that is applied during the *forward* search takes into account the minimum key in the *backward* priority queue.

Nevertheless, there are several examples for bidirectional Dijkstra-like target-oblivious shortest-path algorithms: the bidirectional version of Di-

---

[2]The simplest example of such a criterion allows stopping the search as soon as forward and backward search have settled a common node (i.e., d' drops below ∞). This simple criterion applies to the bidirectional version of Dijkstra's algorithm. For highway hierarchies and highway-node routing, we have to use weaker criteria to ensure correctness (cp. Sections 3.4.2 and 4.3.4).

jkstra's algorithm, highway hierarchies, highway–node routing, the *self-bounding* variant of reach–based routing [26], and the separator–based multi-level method [37]—provided that the abort–on–success criterion is omitted in each case.

5.3   A Generic Algorithm

In a given graph $G = (V, E)$ and for given node sets $S, T \subseteq V$, we want to compute the shortest–path distances $d(s, t)$ for all node pairs $(s, t) \in S \times T$. For this problem, we derive a generic algorithm step–by–step, starting with any bidirectional Dijkstra–like shortest–path algorithm. The final outcome can be found in Figure 5.3(f). All subsequently listed algorithmic variants take source and target node sets S and T as input and compute a distance table D as output such that $D[s, t] = d(s, t)$. Figure 5.1(a) gives a naive bidirectional many–to–many algorithm. Basically, for each s–t–pair, we perform one normal bidirectional query to determine the distance from s to t.

	1	**foreach** $s \in S$ **do**
	2	$\quad$ **foreach** $t \in T$ **do**
(a)	3	$\quad\quad$ compute $\overrightarrow{\sigma}(s, t)$;
	4	$\quad\quad$ compute $\overleftarrow{\sigma}(s, t)$;
	5	$\quad\quad$ compute $D[s, t]$ according to Equation 5.1;

Figure 5.1: Naive many–to–many algorithm based on any bidirectional Dijkstra–like shortest–path algorithm.

Lemma 17 *The naive bidirectional many-to-many algorithm is correct.*

Proof. Follows immediately from Definition 11. $\square$

The first step to get to an efficient many–to–many algorithm is to restrict ourselves to target–oblivious algorithms. For this subclass of bidirectional Dijkstra–like shortest–path algorithms, we can restate our naive algorithm from Figure 5.1(a) in Figure 5.2(b). At first glance, this restriction might seem counter–intuitive since target–oblivious approaches cannot employ an

abort–on–success criterion and, thus, are usually *less* efficient than a cor–
responding 'target–aware' variant. However, now, we can exploit the fact
that the forward search space does not depend on t and the backward search
space does not depend on s, which yields an equivalent version (c) of our al–
gorithm (also given in Figure 5.2). Interestingly, this simple transformation
already constitutes the most important step: instead of performing $|S|$ *times*
$|T|$ bidirectional shortest–path queries (b), we now (c) have to perform only
$|S|$ forward *plus* $|T|$ backward queries (followed by 'intersecting' the search
spaces to compute the actual distances), which is a great improvement.

(b)

1 **foreach** $s \in S$ **do**
2   **foreach** $t \in T$ **do**
3   compute $\overrightarrow{\sigma}(s)$;
4   compute $\overleftarrow{\sigma}(t)$;
5   compute $D[s, t]$ according to Equation 5.1;

$\Updownarrow$

(c)

1 **foreach** $t \in T$ **do** compute $\overleftarrow{\sigma}(t)$;
2 **foreach** $s \in S$ **do**
3   compute $\overrightarrow{\sigma}(s)$;
4   **foreach** $t \in T$ **do**
5   compute $D[s, t]$ according to Equation 5.1;

Figure 5.2: Getting to an efficient many–to–many algorithm based on any
bidirectional Dijkstra–like target–oblivious shortest–path algorithm.

The remaining question is how to evaluate Equation 5.1—i.e., how to
intersect the search spaces—efficiently (Line 5 of Variant (c)). In order to
do so, we associate with each node u a bucket that contains for each target t
whose backward search reaches u an entry consisting of the ID of the target
node and the computed distance y to the target. More formally, we define a
bucket $\beta(u)$ of $u \in V$:

$$\beta(u) := \{(t, y) \mid t \in T \wedge (u, y) \in \overleftarrow{\sigma}(t)\}. \tag{5.3}$$

Now, we can restate Equation 5.1 in the following way:

$$d(s, t) = \min\{x + y \mid (u, x) \in \overrightarrow{\sigma}(s) \wedge (t, y) \in \beta(u)\} \tag{5.4}$$

This equation means that we can compute the optimal distance from s to t by considering for each node u in the forward search space of s, the entry in the bucket of u that matches the target t, adding up the lengths of the paths from s to u and from u to t, and taking the minimum. The correctness is proved in the following lemma.

Lemma 18 *For any bidirectional Dijkstra-like target-oblivious shortest-path algorithm, Equation 5.4 is fulfilled for any node pair* $(s, t) \in V^2$.

Proof. Due to Definition 11, we know that Equation 5.1 is fulfilled for any bidirectional Dijkstra–like shortest–path algorithm. Thus, it is sufficient to show that

$$A := \{x + y \mid (u, x) \in \overrightarrow{\sigma}(s) \wedge (u, y) \in \overleftarrow{\sigma}(t)\} \quad =$$
$$B := \{x + y \mid (u, x) \in \overrightarrow{\sigma}(s) \wedge (t, y) \in \beta(u)\}.$$

This is the case since

$$z \in A$$
$$\Leftrightarrow \quad \exists x, y \in \mathbb{R}_0^+, u \in V : z = x + y \wedge (u, x) \in \overrightarrow{\sigma}(s) \wedge (u, y) \in \overleftarrow{\sigma}(t)$$
$$\overset{(5.3)}{\Leftrightarrow} \quad \exists x, y \in \mathbb{R}_0^+, u \in V : z = x + y \wedge (u, x) \in \overrightarrow{\sigma}(s) \wedge (t, y) \in \beta(u)$$
$$\Leftrightarrow \quad z \in B. \qquad \qquad \qquad \qquad \qquad \qquad \qquad \qquad \qquad \square$$

We obtain a new algorithmic variant (Figure 5.3(d)) that is equivalent to (c), but employs Equation 5.4 instead of 5.1, as explicitly specified in Lines 7–9. Note that this requires a proper initialisation of the distance table (Line 1) and the composition of the buckets (Line 3).

Now, we can swap Lines 6 and 7 (since $\overrightarrow{\sigma}(s)$ does not depend on t), yielding the equivalent variant (e). Finally, we can merge[3] Lines 7 and 8, resulting in the final, very efficient variant (f).

Theorem 7 *The many-to-many algorithm (Figure 5.3(f)) is correct.*

Proof. Follows directly from Lemma 17 and the fact that the algorithmic variants (a), (b), (c), (d), (e), and (f) are equivalent. $\qquad \square$

---

[3] At first glance, it seems that Line (e)–7 just disappears. However, it is important to note that in Line (f)–7, t is no longer bound (as in Line (e)–8), but free.

```
    1  foreach (s, t) ∈ S × T do D[s, t] := ∞;
    2  foreach t ∈ T do compute σ⃖(t);
    3  foreach u ∈ V do compose β(u) according to Equation 5.3;
    4  foreach s ∈ S do
    5      compute σ⃗(s);
```

(d)
```
    6      foreach t ∈ T do
    7          foreach (u, x) ∈ σ⃗(s) do
    8              foreach (t, y) ∈ β(u) do
    9                  D[s, t] := min(D[s, t], x + y);
```

$$\Updownarrow$$

(e)
```
    6      foreach (u, x) ∈ σ⃗(s) do
    7          foreach t ∈ T do
    8              foreach (t, y) ∈ β(u) do
    9                  D[s, t] := min(D[s, t], x + y);
```

$$\Updownarrow$$

(f)
```
    6      foreach (u, x) ∈ σ⃗(s) do
    7          foreach (t, y) ∈ β(u) do
    8              D[s, t] := min(D[s, t], x + y);
```

Figure 5.3: Getting to an efficient many-to-many algorithm based on any bidirectional Dijkstra-like target-oblivious shortest-path algorithm (continued from Figure 5.2). Note that the three variants (d), (e), and (f) have the first five lines in common. Variant (f) is the final variant.

5.4   A Concrete Instantiation

The generic many-to-many algorithm (Figure 5.3(f)) can be instantiated based on any bidirectional Dijkstra-like target-oblivious shortest-path algorithm. However, not every instantiation is reasonable. For example, using the bidirectional version of Dijkstra's algorithm without abort-on-success criterion implies that both backward and forward search scan the *complete* graph. This is by far worse than just employing $|S|$ unidirectional Dijkstra searches.

In contrast, a direct application of highway-node routing (without abort-on-success criterion)[4] is very promising since the search spaces are very small. The same applies to highway hierarchies. The subsequent descriptions refer to the many-to-many algorithm based on highway-node routing, knowing that an instantiation based on highway hierarchies can be achieved in an analogous way. (At some point in Section 5.4.2, we will distinguish between highway-node routing and highway hierarchies to introduce optimisations that are specific to each particular instantiation.)

When considering the search spaces of highway-node routing, we observe that both forward and backward search typically scan all nodes of the topmost overlay graph. This causes unnecessary efforts that can be avoided by introducing an asymmetric variant of highway-node routing.

5.4.1   Asymmetry

Without loss of generality, we assume $|T| \geq |S|$. Otherwise, it is more efficient to apply our algorithm to the reverse graph.

We specify an *asymmetric highway-node routing* algorithm that is a very simple modification of the original highway-node routing query algorithm: the backward search is pruned at nodes $u \in V_L$, i.e., outgoing edges of such nodes u are not relaxed.

Lemma 19 *The asymmetric variant of highway-node routing is a bidirectional Dijkstra-like target-oblivious shortest-path algorithm.*

---

[4]In the context of the many-to-many algorithm based on a *target-oblivious* shortest-path algorithm, we always refer to approaches that do not employ an abort-on-success criterion. Thus, subsequently, we will no longer state this explicitly each time.

Proof. Consider the proof of Theorem 6 ("correctness of highway-node routing"). We have shown that there is a particular meeting point p on a shortest path P such that (i) $(p, d(s, p)) \in \overrightarrow{\sigma}(s)$ and (ii) $(p, d(p, t)) \in \overleftarrow{\sigma}(t)$. This still holds since (i) the forward search has not been changed at all and (ii) the meeting point p has been defined in such a way that all nodes $u \neq p$ on $P|_{p \to t}$ do not belong to V_L so that the backward search is not pruned at these nodes u, which implies that p is settled. We can conclude that Equation 5.1 is fulfilled. Moreover, it is obvious that Equation 5.2 is fulfilled as well. □

From this lemma, we can conclude that the application of the asymmetric variant of highway-node routing yields a correct (and more efficient) many-to-many algorithm. Note that not only the backward search space sizes have been reduced, but also the total number of bucket entries and, thus, the number of bucket scans during the forward searches.

Analysis. Since highway-node routing does not give worst case performance guarantees that hold for arbitrary graphs, our analysis will be based on parameterisations and assumptions that still have to be checked experimentally. We nevertheless believe that such an analysis is valuable because it explains the behaviour of the algorithm and helps finding even better variants.

In the following, we regard L as being a tuning parameter, i.e., we allow the choice of different numbers of levels of the multi-level overlay graph. We write $\overrightarrow{\sigma}(s, L)$ and $\overleftarrow{\sigma}(t, L)$ (instead of $\overrightarrow{\sigma}(s)$ and $\overleftarrow{\sigma}(t)$) to stress that the forward and backward search spaces depend on the parameter L. It will be more convenient to disregard the associated distances and consider only the nodes in the search spaces. For this purpose, we introduce

$$\overrightarrow{\sigma}_V(s, L) := \{v \mid (v, x) \in \overrightarrow{\sigma}(s, L)\} \quad \text{and}$$
$$\overleftarrow{\sigma}_V(t, L) := \{v \mid (v, x) \in \overleftarrow{\sigma}(t, L)\}.$$

Let $X(L)$ denote the average size of the backward search spaces, i.e.,

$$X(L) := \left(\sum_{t \in T} |\overleftarrow{\sigma}_V(t, L)| \right) / |T|.$$

Note that a forward search explores roughly $X(L) + |V_L|$ nodes on average since it searches up to the topmost level (visiting roughly the same number of nodes as a backward search) and continues in the topmost level (visiting $|V_L|$ nodes). Moreover, let $Y(L)$ denote the overlap ratio of forward and backward searches, i.e.,

$$Y(L) := \left(\frac{\sum_{s \in S, t \in T} |\overrightarrow{\sigma}_V(s, L) \cap \overleftarrow{\sigma}_V(t, L)|}{|S| \cdot |T|} \right) / X(L),$$

and let $T_{\text{Dijk}}(k)$ denote the cost of a Dijkstra-like search when exploring k nodes in a road network.

The backward searches have cost $|T| \cdot T_{\text{Dijk}}(X(L))$. Building buckets costs time $O(|T| \cdot X(L))$, which is dominated by the time required for the backward searches so that we need not explicitly regard this term in our total cost calculation. The forward searches have cost about $|S| \cdot T_{\text{Dijk}}(X(L) + |V_L|)$ for the search itself. Bucket scanning takes

$$O\left(\sum_{s \in S, t \in T} |\overrightarrow{\sigma}_V(s, L) \cap \overleftarrow{\sigma}_V(t, L)| \right) = O\Big(|S| \cdot |T| \cdot Y(L) \cdot X(L)\Big).$$

We get a total cost of

$$\underbrace{|S| \cdot T_{\text{Dijk}}(X(L) + |V_L|)}_{1.} + \underbrace{|T| \cdot T_{\text{Dijk}}(X(L))}_{2.} + \underbrace{O(|S| \cdot |T| \cdot Y(L) \cdot X(L))}_{3.}.$$

Note that the first term clearly *decreases* for an increasing L (since the decrease of $|V_L|$ outweighs the increase of $X(L)$), while the second term grows for an increasing L. The third term is less obvious since the characteristics of $Y(L)$ are less clear. Our experiments (see Table 7.19 in Section 7.8) indicate that $Y(L)$ fluctuates within a small range in case of highway-node routing and increases (for an increasing L) in case of highway hierarchies. We can conclude that typically the third term grows with L since $X(L)$ clearly grows.

If both sets S and T are large, the third term dominates. From this we can learn two things. First, since the constant behind this term is very small, we can expect very good performance for large problems. Second, we can actually save time by choosing L smaller than the maximum possible level. We could use random sampling to estimate the amount of overlap present in the input. Based on this estimate and appropriately measured constants of proportionality, we would then get a cost model that is accurate enough to choose a (near) optimal value for L.

It is also interesting to look at extreme cases. When $|S| = |T| = 1$, it is best to choose the highest possible level as L and we essentially get the asymmetric variant of point–to–point highway–node routing. When $T = V$, it is best to choose $L = 0$ and we just get Dijkstra's algorithm for re–peatingly solving the single–source shortest–path problem from all nodes $s \in S$. In other words, our many–to–many algorithm smoothly interpolates between good algorithms for these extreme cases and promises considerable speedups in the middle where none of the 'extreme' algorithms works very well.

5.4.2   Optimisations

Fewer Bucket Entries.   We can speed up bucket scanning by reducing the number of bucket entries that are made during the backward searches. For determining the correct shortest path length $d(s, t)$, it is sufficient that there is a single bucket entry $(t, d(u, t))$ at some node u on a shortest s–t–path such that $(u, d(s, u)) \in \overrightarrow{\sigma}(s)$. Every additional bucket entry with this property costs unnecessary extra scanning time.

In case of *highway hierarchies*, we observe that during a backward search the current search level can differ from the actual level of a node. Due to this fact, bucket entries at nodes in the core of level L are made while the search is still in a level $\ell < L$. Because every forward search settles all nodes in G'_L, a bucket entry $(t, d) \in \beta(u)$ can be omitted if u has been settled on a path $\langle u, \dots, v, \dots, t \rangle$ such that both u and v are in the core of level L.

This observation does not apply to *highway-node routing*, where we do not distinguish between search level and node level. However, in case of highway–node routing, we can reduce the number of bucket entries by ignoring nodes u that have been *stalled*: if we know that the computed distance d from u to the current target t is suboptimal, we certainly need not store a bucket entry (t, d) at u. We will see that by this means, the number of bucket entries can be reduced by about 40% (cp. Table 7.19 in Section 7.8).

Accurate Backward Search. The reduction of bucket scans described in the previous paragraph can be strengthened by performing accurate backward searches. The current version of backward search is not accurate because we break the search when entering (the core of) the topmost level. In a sense, this corresponds to the 'aggressive approach' introduced in Section 4.2.4.

In case of *highway hierarchies*, we can think of continuing the backward search until all nodes in the priority queue are in the core of the topmost level instead of pruning the search at level–L core entrance points. This method, which corresponds to the 'conservative approach' (Section 4.2.3), leads to fewer bucket entries because the restriction of the previous paragraph applies more often.

In case of *highway-node routing*, we could do similar things. Alternatively, we can at least relax the edges of the pruned nodes in order to wake adjacent nodes so that they can possibly start a stalling process ('stall-on-demand', Section 4.2.6).

Note that for small distance tables these measures might be counterproductive since the backward searches get more expensive.

5.4.3 Algorithmic Details

To deal efficiently with a large amount of buckets with different sizes (which are not known in advance), we suggest the following approach: During the backward searches, we manage a single resizable array representing the set $\{(u, t, d) \mid t \in T \wedge (u, d) \in \overleftarrow{\sigma}(t)\}$. After all backward searches have been performed, we group these triples by the first component. We build an appropriate index structure so that we can access in constant time for any node $u \in V$ the consecutive range of the array that represents $\beta(u)$.

5.4.4 Extensions

Outputting Complete Path Descriptions. So far we have only described how to compute shortest-path distances. We now explain how the algorithm can be modified so that it computes a data structure that allows outputting a complete description of a shortest s–t–path P (for $(s, t) \in S \times T$) in time $O(|P|)$.

We explicitly store the search spaces of forward and backward searches in the form of rooted trees. For each query pair (s,t), there is some intermediate node v such that a shortest path from s to t is composed of an s-v-path in the forward search space from s and a v-t-path in the backward search space to t. Hence, all we need to store are pointers to v in the two search spaces. This information is updated during the main computation whenever $D[s,t]$ is improved. By this means, we can easily assemble a path in the multi-level overlay graph. Then, we can use the techniques introduced in Section 3.4.3 to expand edges in the overlay graphs in order to reconstruct the represented subpaths in the original graph.

We can save some space by pruning those parts of the search spaces that are not needed for any shortest connection. In the case of a forward search space, this pruning can be done directly after the search has finished.

Computing Shortest Connections Incrementally. In some applications, we are not really interested in a complete distance table. For example, many heuristics for the travelling salesman problem start with the closest connections for each node and only compute additional connections on demand [33]. For such applications, the asymmetry in our search algorithm is again helpful. As before, the cheap backward search is done for all nodes $t \in T$. The comparatively expensive forward searches, however, which require heavy scanning of buckets, are only progressing incrementally after their search frontier is completely in the topmost level.

To do this, we remember the number of level-L nodes encountered by each backward search. Each forward search is equipped with a copy of this counter array. When the forward search scans a bucket entry (t,d) of some level-L node, it decrements the counter for t. When the counter reaches zero, we know that $D[s,t] = d(s,t)$ and we can output the newly found distance.

Parallelisation. Suppose we have a shared memory parallel computer with x processing elements (PEs). Then the problem is easy to parallelise: each PE performs $\lceil |T|/x \rceil$ backward searches and $\lceil |S|/x \rceil$ forward searches. If $x > |S|$, we can achieve further parallelism by partitioning $|T|$ and the corresponding buckets into k groups. Now x/k processors are assigned to each group and perform a forward search from all nodes in $|S|$ considering only the target nodes in their group.

5.5 Concluding Remarks

Review. Our algorithm provides a straightforward and highly efficient solution to the many-to-many shortest-path problem. Furthermore, it is generic, i.e., it can be instantiated based on various point-to-point algorithms. This makes it comparatively easy to incorporate our approach into different environments. Moreover, the memory overhead is quite small. When we use our instantiation based on highway-node routing, we also can take advantage of the fact that highway-node routing can handle certain dynamic scenarios. For example, we can recompute the multi-level overlay graph in order to solve a many-to-many instance based on a different cost function.

Our many-to-many algorithm can be employed directly in various real-world applications. Furthermore, it can be used in the preprocessing stage of some point-to-point shortest-path algorithms, namely for precomputed cluster distances (Section 1.2.2) and transit-node routing (next chapter).

Future Work. Let us consider the case that we want to compute only a single table in a given road network. If the table is big enough, our approach beats Dijkstra's algorithm even if the preprocessing of highway-node routing is considered to be part of our task (cp. Section 7.8.2). However, it might be interesting to find ways to compute a multi-level overlay graph specially tailored to S and T—after all we only need to preserve shortest paths between nodes in S and T. The hope would be that this can be done more efficiently than building a complete multi-level overlay graph.

References. This chapter is partly based on [51], which is, in turn, based on Sebastian Knopp's diploma thesis [50]. While the presentation of the algorithm and the experiments in [51] refer to an instantiation based on highway hierarchies, in this thesis, we prefer to introduce the many-to-many algorithm in a generic way. Furthermore, we somewhat improve the analysis of the algorithm. For performing experiments, we employ a new reimplementation based on highway hierarchies and a new implementation based

on highway-node routing. In this thesis, we restrict ourselves to a few randomly generated symmetric (i.e., $S = T$) instances. Further experiments
with randomly generated asymmetric instances and real-world instances can
be found in [51] and [50].

6

Transit-Node Routing

6.1 Central Ideas

Transit-node routing is based on a simple observation intuitively used by humans: When you start from a source node s and drive to somewhere 'far away', you will leave your current location via one of only a few 'important' traffic junctions, called (forward) *access nodes* $\overrightarrow{A}(s)$. An analogous argument applies to the target t, i.e., the target is reached from one of only a few backward access nodes $\overleftarrow{A}(t)$. Moreover, the union of all forward and backward access nodes of all nodes, called *transit-node set* T, is rather small. This implies that for each node the distances to/from its forward/backward access nodes and for each transit-node pair (u, v) the distance between u and v can be stored. For given source and target nodes s and t, the length of the shortest path that passes at least one transit node is given by

$$d_T(s, t) = \min\{d(s, u) + d(u, v) + d(v, t) \mid u \in \overrightarrow{A}(s), v \in \overleftarrow{A}(t)\}.$$

Note that all involved distances $d(s, u)$, $d(u, v)$, and $d(v, t)$ can be directly looked up in the precomputed data structures. As a final ingredient, a *locality filter* $\mathcal{L} : V \times V \rightarrow \{\text{true}, \text{false}\}$ is needed that decides whether given nodes s and t are too close to travel via a transit node. $\mathcal{L}$ has to fulfil the property that $\neg\mathcal{L}(s, t)$ implies $d(s, t) = d_T(s, t)$. Note that in general the converse need not hold since this might hinder an efficient realisation of the locality filter. Thus, *false positives*, i.e., "$\mathcal{L}(s, t) \wedge d(s, t) = d_T(s, t)$", may occur.

The following algorithm can be used to compute $d(s, t)$:

1   **if** $\neg\mathcal{L}(s,t)$ **then** compute and return $d_T(s,t)$;
2   **else** use any other routing algorithm.

Figure 6.1 gives a *schematic* representation of transit–node routing, while
Figure 6.2 (first published in [5]) gives a *real-world* example.

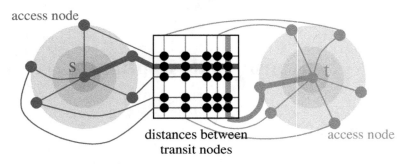

Figure 6.1: Schematic representation of transit–node routing.

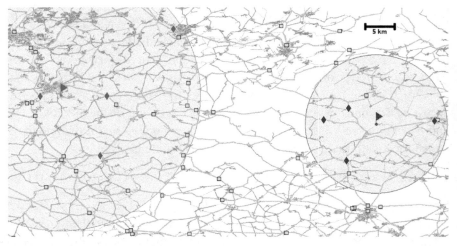

Figure 6.2: Finding the optimal travel time between two points (flags) some–
where between Saarbrücken and Karlsruhe amounts to retrieving the two
times four *access nodes* (diamonds), performing 16 table lookups between
all pairs of access nodes, and checking that the two disks defining the *locality filter* do not overlap. *Transit nodes* that do not belong to the access–node
sets of the selected source and target nodes are drawn as small squares.

Knowing the length of the shortest path, a complete description of it can be efficiently derived using iterative table lookups and precomputed rep-resentations of paths between transit nodes. Provided that the above ob-servation holds and that the percentage of false positives is low, the above algorithm is very efficient since a large fraction of all queries can be han-dled in Line 1, $d_T(s,t)$ can be computed using only a few table lookups, and source and target of the remaining queries in Line 2 are quite close. In fact, the remaining queries can be further accelerated by introducing additional levels of transit-node routing.

6.2 A Generic Algorithm

For a given graph $G = (V, E)$, we consider $L + 1$ sets

$$V =: \mathcal{T}_0 \supseteq \mathcal{T}_1 \supseteq \cdots \supseteq \mathcal{T}_L$$

of *transit nodes*.[1] Moreover, for any level $\ell, 0 \leq \ell \leq L$, we consider

- a forward and a backward *access mapping* $\overrightarrow{A}_\ell : V \to 2^{\mathcal{T}_\ell}$ and $\overleftarrow{A}_\ell : V \to 2^{\mathcal{T}_\ell}$, which map a node to its forward and backward *access nodes*, respectively,

- a *locality filter* $\mathcal{L}_\ell : V \times V \to \{\text{true}, \text{false}\}$, which decides whether the distance between two nodes can be determined using only levels $\geq \ell$ of transit-node routing,

- a distance table $D_\ell : \mathcal{T}_\ell \times \mathcal{T}_\ell \to \mathbb{R}_0^+ \cup \{\infty\}$, which contains the correct distances between all node pairs from $\mathcal{T}_\ell \times \mathcal{T}_\ell$ except for the distances that can be computed using higher levels of transit-node routing,

- the distance $d_\ell : V \times V \to \mathbb{R}_0^+ \cup \{\infty\}$ that is obtained using level ℓ of transit-node routing, i.e., considering all access nodes to level ℓ and the distances between all pairs of these access nodes, and

- the minimal distance $d_{\geq \ell} : V \times V \to \mathbb{R}_0^+ \cup \{\infty\}$ that can be obtained using all levels $\geq \ell$.

---

[1]Note that in earlier publications [71, 4, 5], the order of the levels (which we called 'layers' at that time) was reversed: the topmost transit-node set was denoted by $\mathcal{T}_1$, now it is denoted by $\mathcal{T}_L$. We have changed the order to blend well with the other hierarchical approaches presented in this thesis.

To avoid some case distinctions, we introduce the following definitions:

- $\mathcal{T}_{L+1} := \emptyset$,
- $\overrightarrow{A}_0(u) := \overleftarrow{A}_0(u) := \{u\}$,
- $d_{\geq L+1}(s,t) := \infty$,
- $\min \emptyset := \infty$.

Now, for any level $\ell, 0 \leq \ell \leq L$, we give a precise definition of the three distance functions D_ℓ, d_ℓ, and $d_{\geq \ell}$:

$$D_\ell(s,t) := \begin{cases} d(s,t) & \text{if } d(s,t) < d_{\geq \ell+1}(s,t) \\ \infty & \text{otherwise} \end{cases} \tag{6.1}$$

$$d_\ell(s,t) := \min\{d(s,u)+D_\ell(u,v)+d(v,t) \mid u \in \overrightarrow{A}_\ell(s), v \in \overleftarrow{A}_\ell(t)\} \tag{6.2}$$

$$d_{\geq \ell}(s,t) := \min_{k \geq \ell} d_k(s,t) \tag{6.3}$$

Note that the following equation is equivalent to (6.3):

$$d_{\geq \ell}(s,t) = \min(d_\ell(s,t), \min_{k \geq \ell+1} d_k(s,t)) \tag{6.4}$$

Obviously, all these distances are upper bounds on the actual shortest-path length, as stated in the following proposition:

Proposition 5 $D_\ell(s,t) \geq d(s,t)$, $d_\ell(s,t) \geq d(s,t)$, $d_{\geq \ell}(s,t) \geq d(s,t)$.

We assume that all distances to/from forward/backward access nodes and all distances $D_\ell(s,t)$ have been precomputed. We can show that we always obtain the correct shortest-path length when we use all levels of transit-node routing:

Lemma 20 $d_{\geq 0}(s,t) = d(s,t)$.

Proof. Due to (6.2), we have $d_0(s,t) = d(s,s) + D_0(s,t) + d(t,t)$ since $\overrightarrow{A}_0(s) = \{s\}$ and $\overleftarrow{A}_0(t) = \{t\}$. If $d(s,t) < d_{\geq 1}(s,t)$, we have $d_0(s,t) = D_0(s,t) = d(s,t)$ (due to (6.1)) and thus, $d_{\geq 0}(s,t) = \min(d_0(s,t), d_{\geq 1}(s,t)) = d(s,t)$ by (6.4) and Proposition 5. Otherwise $(d(s,t) = d_{\geq 1}(s,t))$, we have $d_0(s,t) = D_0(s,t) = \infty$ (due to (6.1)) and, again, $d_{\geq 0}(s,t) = \min(d_0(s,t), d_{\geq 1}(s,t)) = d(s,t)$. $\qquad\square$

Of course, using all levels is comparatively expensive. Therefore, we want to avoid accessing levels that are not needed to get the correct result. For the decision making we want to employ the already introduced locality filters. We require that

$$\neg \mathcal{L}_\ell(s,t) \rightarrow (d(s,t) = d_{\geq \ell}(s,t)). \qquad (6.5)$$

Then, we can use the *transit-node routing* algorithm as specified in Figure 6.3 to efficiently compute the length of a shortest path from a given source node s to a given target node t.

input: source node s and target node t
output: distance $d(s,t)$

```
1  d' := ∞;
2  for ℓ := L downto 0 do
3       d' := min(d', dℓ(s,t));
4       assert d' = d≥ℓ(s,t);
5       if ¬Lℓ(s,t) then break;
6  return d';
```

Figure 6.3: The transit-node routing algorithm.

Theorem 8 *Transit-node routing is correct.*

Proof. If the condition in Line 5 is fulfilled at some point, we return $d' = d_{\geq \ell}(s,t) = d(s,t)$ due to (6.5). Otherwise, we return $d' = d_{\geq 0}(s,t) = d(s,t)$ according to Lemma 20. □

Practical Remarks. In the distance tables D_ℓ, it is sufficient to store only the non-infinity entries explicitly. For this purpose, we can use a space-efficient static hash table. Furthermore, as an alternative to precomputing the entries in D_0, we can use any other shortest-path algorithm to compute the distances D_0 on-the-fly when they are required.

6.3 An Abstract Instantiation

In this section, we instantiate the algorithm from the previous section by giving concrete access mappings, while a concrete choice of the transit–node sets is still *not* specified. The locality filter will be defined in such a way that Equation 6.5 is fulfilled. Note that other instantiations of the generic algorithm that deviate from this section are possible (cp. Section 6.5: "Alternative Instantiations").

Access Mapping. For a node s and a level ℓ, consider a set C of covering paths[2] of s w.r.t. $\mathcal{T}_\ell$ in G. (To obtain a very efficient algorithm, we might want to choose a *minimal* covering–paths set.) Let

$$\overrightarrow{A}_\ell(s) := \{v \mid P = \langle s, \ldots, v \rangle \in C\}.$$

The backward access mapping is defined analogously, considering the reverse graph $\overleftarrow{G}$ instead of G.

Locality Filter. An explicit representation of a level–ℓ locality filter (storing n^2 bits) would need too much space for large graphs. Therefore, we look for a more space–efficient alternative. We want to identify node pairs (s, t) such that the distance $d(s, t)$ cannot be computed using transit–node routing in level ℓ or higher. For each of these pairs, we pick one *witness*, a particular node p on a shortest s–t–path. We make sure that both s and t memorise this witness p. Then, when we want to evaluate $\mathcal{L}_\ell(s, t)$, we just have to check whether s and t share a common witness. Note that this approach can lead to *false positives*, i.e., two nodes might share a common witness although their distance actually can be computed using transit–node routing in level ℓ or higher.

   Beside paying attention to the memory requirements, we are also interested in fast preprocessing times. Therefore, we introduce the concept of *handing* computed data *down* from higher to lower levels: Let us consider some path $\langle s_0, \ldots, s_1, \ldots, p, \ldots, t_1, \ldots, t_0 \rangle$ with $s_0, t_0 \in \mathcal{T}_0$ and $s_1, t_1 \in \mathcal{T}_1$. Moreover, let us assume that we already know that $d(s_1, t_1)$

---

[2]The concept of *covering paths* has been introduced in Section 4.2. Note that in contrast to Chapter 4, here we do *not* require *canonical* covering paths.

cannot be computed using level 2 or higher. Thus, we have some witness p and both s_1 and t_1 memorise this witness. Now, this witness is handed down from s_1 to s_0 and from t_1 to t_0. An equivalent formulation is to say that s_0 inherits the witness p from s_1. Now, if we want to decide whether $d(s_0, t_0)$ can be determined using level 2 or higher, the answer is 'no' since s_0 and t_0 share the common witness p. Note that by this means, the number of false positives may increase.

In the following, we work out the formal details of these ideas. The *level* $\ell(u)$ of a node $u \in V$ is $\max\{\ell \mid u \in T_\ell\}$. Let us assume that we have some fixed strategy that picks for any two connected nodes s and t one particular node $p(s,t)$ on one particular shortest s-t-path. We define forward and backward node sets $\overrightarrow{K}_\ell : V \to 2^V$ and $\overleftarrow{K}_\ell : V \to 2^V$ in the following way: for any node s and any level $\ell < \ell(s) + 1$, $\overrightarrow{K}_\ell(s) := \emptyset$, for level $\ell = \ell(s) + 1$,

$$\overrightarrow{K}_\ell(s) := \{p(s,t) \mid t \in V \wedge \ell(s) = \ell(t) \wedge d(s,t) < d_{\geq \ell}(s,t)\} \qquad (6.6)$$

and for any level $\ell > \ell(s) + 1$,

$$\overrightarrow{K}_\ell(s) := \bigcup_{u \in \overrightarrow{A}_{\ell-1}(s)} \overrightarrow{K}_\ell(u), \qquad (6.7)$$

and analogously, for any node t and any level $\ell < \ell(t) + 1$, $\overleftarrow{K}_\ell(t) := \emptyset$, for level $\ell = \ell(t) + 1$,

$$\overleftarrow{K}_\ell(t) := \{p(s,t) \mid s \in V \wedge \ell(s) = \ell(t) \wedge d(s,t) < d_{\geq \ell}(s,t)\} \qquad (6.8)$$

and for any level $\ell > \ell(t) + 1$,

$$\overleftarrow{K}_\ell(t) := \bigcup_{u \in \overleftarrow{A}_{\ell-1}(t)} \overleftarrow{K}_\ell(u). \qquad (6.9)$$

Note that Equations 6.6 and 6.8 reflect the 'witness' idea, while Equations 6.7 and 6.9 reflect the 'handing down' idea.

Finally, we define the locality filter

$$\mathcal{L}_\ell(s,t) := \bigvee_{k \leq \ell} \left(\overrightarrow{K}_k(s) \cap \overleftarrow{K}_k(t) \neq \emptyset \right). \qquad (6.10)$$

Lemma 21 *Consider two nodes s and t with $d(s,t) \neq \infty$. If and only if there is some node $u \in \mathcal{T}_\ell$ on some shortest s-t-path P, then $d_{\geq\ell}(s,t) = d(s,t)$.*

Proof. $\Leftarrow$) We have $d_{\geq\ell}(s,t) = d(s,t)$. This implies, by (6.3) and (6.2), that there is a level $k \geq \ell$, a node $u \in \vec{A}_k(s)$, and a node $v \in \overleftarrow{A}_k(t)$ such that $d(s,u) + D_k(u,v) + d(v,t) = d(s,t)$. Due to Proposition 5, we have $D_k(u,v) \geq d(u,v)$. We can conclude that u and v are nodes on a shortest s-t-path. Furthermore, we know that $u \in \vec{A}_k(s) \subseteq \mathcal{T}_k \subseteq \mathcal{T}_\ell$.

$\Rightarrow$) We pick the maximum level $k \geq \ell$ with the property that there is some node from $\mathcal{T}_k$ on some shortest s-t-path P. Let u and v denote the first and the last node from $\mathcal{T}_k$ on P, respectively. The case $u = v$ is possible. According to the definitions of the covering paths and the access mappings, there is a node $u' \in \vec{A}_k(s) \subseteq \mathcal{T}_k$ on a shortest s-u-path $\vec{P}$ and a node $v' \in \overleftarrow{A}_k(t) \subseteq \mathcal{T}_k$ on a shortest v-t-path $\overleftarrow{P}$. Consider the path

$$P' := \langle s, \ldots, \overbrace{u', \ldots, \underbrace{u, \ldots, v}_{P|_{u \to v}}, \ldots, v'}^{\overleftarrow{P}}, \ldots, t \rangle,$$

which is a shortest s-t-path as well. According to (6.2), we have $d_k(s,t) \leq d(s,u') + D_k(u',v') + d(v',t)$. Due to our choice of k, we know that there is no node $x \in \mathcal{T}_{k+1}$ on any shortest u'-v'-path Q—otherwise, the same node x would be on the shortest s-t-path $P'|_{s \to u'} \circ Q \circ P'|_{v' \to t}$. From the part of this lemma that has already been proven, it follows that $d(u',v') < d_{\geq k+1}(u',v')$. Thus, by (6.1), $D_k(u',v') = d(u',v')$ and, consequently, $d_k(s,t) \leq d(s,u') + D_k(u',v') + d(v',t) = d(s,t)$. From $d_{\geq\ell}(s,t) \leq d_k(s,t) \leq d(s,t)$, it follows that $d_{\geq\ell}(s,t) = d(s,t)$ due to Proposition 5. $\square$

Lemma 22 *The locality filter specified in Equation 6.10 fulfils Equation 6.5.*

Proof. Trivial for $d(s,t) = \infty$ (due to Proposition 5). For $d(s,t) \neq \infty$, we want to show the contraposition of Equation 6.5 and therefore assume that $d(s,t) \neq d_{\geq\ell}(s,t)$. Let k be the maximum level such that $d_{\geq k-1}(s,t) = d(s,t)$. Such a k must exist due to Lemma 20. The choice of k implies $k - 1 < \ell$, $d_{\geq k}(s,t) \neq d(s,t)$, and $d_{k-1}(s,t) = d(s,t)$. Hence, there is some shortest s-t-path with nodes $u' \in \overrightarrow{A}_{k-1}(s)$ and $v' \in \overleftarrow{A}_{k-1}(t)$ on it. If $s \in \mathcal{T}_{k-1}$, we set $u := s$; otherwise, $u := u'$. Analogously, if $t \in \mathcal{T}_{k-1}$, we set $v := t$; otherwise, $v := v'$. In any case, we have $u, v \in \mathcal{T}_{k-1}$.

Lemma 21 and $d_{\geq k}(s,t) \neq d(s,t)$ imply that there is no shortest s-t-path that contains a node from $\mathcal{T}_k$. In particular, $u, v \notin \mathcal{T}_k$ and $d(u,v) < d_{\geq k}(u,v)$—otherwise, there would be a shortest u-v-path containing a node $x \in \mathcal{T}_k$ and thus, also a shortest s-t-path containing x. Since $u, v \in \mathcal{T}_{k-1} \setminus \mathcal{T}_k$, we have $\ell(u) = \ell(v) = k - 1$. We can conclude that $p(u,v) \in \overrightarrow{K}_k(u) \cap \overleftarrow{K}_k(v)$ due to Equations 6.6 and 6.8. If $s \notin \mathcal{T}_{k-1}$, we have $\ell(s) < k - 1$, which implies $p(u,v) \in \overrightarrow{K}_k(s)$ due to Equation 6.7: s inherits $p(u,v)$ from $u = u'$. Otherwise, we have $s = u$ so that $p(u,v) \in \overrightarrow{K}_k(s)$ holds as well. An analogous argument applies to $\overleftarrow{K}_k(t)$. Thus, $p(u,v) \in \overrightarrow{K}_k(s) \cap \overleftarrow{K}_k(t)$. Since $k \leq \ell$, $\mathcal{L}_\ell(s,t) =$ true according to (6.10). □

6.3.1 Computing Access Nodes

Here, we describe how to determine the forward access nodes to the topmost level L. Analogous methods can be applied to compute forward and backward access nodes to different levels. From each node $u \in V$, we perform a local Dijkstra search in G in order to determine the covering-paths set w.r.t. $\mathcal{T}_L$ (cp. Section 4.2.2). We take each endpoint of a covering path as access node of u. Applied naively, this approach is rather inefficient. However, we can use two tricks to make it efficient.

First, we do not have to use the conservative approach (Section 4.2.3), but we can use one of the more advanced techniques (Sections 4.2.4–4.2.6). However, in general, these techniques do not yield a *minimal* access-node set, which would be preferable. Fortunately, the resulting set can be easily *reduced* if the distances between all transit nodes are already known: if an access node y can be reached from u via another access node w on a shortest path, we can discard y. Figure 6.4 gives an example. Note that this reduction procedure is very similar to the edge reduction in Section 4.3.3.

Lemma 23 *Applying the reduction procedure yields a minimal access-node set.*

We omit a detailed proof here since it would be very similar to the proof of Lemma 16.

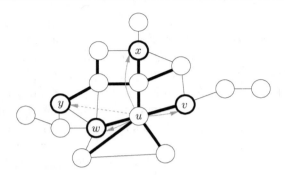

Figure 6.4: Example for the computation of access nodes including the first, but not the second 'trick'. Edge weights correspond to the lengths of the drawn line segments. The nodes v, w, x, and y belong to $\mathcal{T}_L$. The search is started from u. All thick edges belong to the search tree. All depicted nodes from $\mathcal{T}_L$ are endpoints of covering paths. However, y can be removed from this set since the path from u via w to y turns out to be shorter than the path that has been found. Thus, u has only three access nodes.

Second, we may only determine the access node sets $\overrightarrow{A}_L(v)$ for all nodes $v \in \mathcal{T}_{L-1}$ and the sets $\overrightarrow{A}_{L-1}(u)$ for all nodes $u \in V$. Then, for each node $u \in V$, we can compute

$$\overrightarrow{A}_L(u) := \bigcup_{v \in \overrightarrow{A}_{L-1}(u)} \overrightarrow{A}_L(v).$$

Again, we can use the reduction technique to remove unnecessary elements from the set union. The idea to hand access nodes down can be extended to work across more than one level:

$$\overrightarrow{A}_L(u_0) := \bigcup_{u_1 \in \overrightarrow{A}_1(u_0)} \bigcup_{u_2 \in \overrightarrow{A}_2(u_1)} \cdots \bigcup_{u_{L-1} \in \overrightarrow{A}_{L-1}(u_{L-2})} \overrightarrow{A}_L(u_{L-1}).$$

$$(6.11)$$

Lemma 24 *Handing down access nodes is correct, i.e., the resulting access-node set complies with the specification at the beginning of Section 6.3.*

Proof. We say that an access-node set $\overrightarrow{A}_\ell(u)$ is *proper* (i.e., it complies with the specification at the beginning of Section 6.3) iff there is a covering-paths set $C_\ell(u)$ of u w.r.t. $\mathcal{T}_\ell$ such that $\overrightarrow{A}_\ell(u) = \{v \mid P = \langle u, \ldots, v\rangle \in C_\ell(u)\}$.

Assume that for some node u and some level $\ell > 0$, we have a proper access-node set $\overrightarrow{A}_{\ell-1}(u)$ (and thus, a corresponding covering-paths set $C_{\ell-1}(u)$) and that for each node $v \in \overrightarrow{A}_{\ell-1}(u)$, we have a proper access-node set $\overrightarrow{A}_\ell(v)$ (and thus, a corresponding covering-paths set $C_\ell(v)$). Let

$$\overrightarrow{A}_\ell(u) := \bigcup_{v \in \overrightarrow{A}_{\ell-1}(u)} \overrightarrow{A}_\ell(v)$$

and

$$C_\ell(u) := \{P = \langle u, \ldots, v\rangle \mid P \in \mathcal{U}(G) \wedge v \in \overrightarrow{A}_\ell(u)\}.$$

We have to prove that $\overrightarrow{A}_\ell(u)$ is a proper access-node set. For that, it is sufficient to show that $C_\ell(u)$ is a covering-paths set of u w.r.t. $\mathcal{T}_\ell$.

Consider any node $t \in \mathcal{T}_\ell$ that can be reached from u. We have to show that there is a node $x \in \mathcal{T}_\ell$ on some shortest u-t-path P such that $P|_{u \to x} \in C_\ell(u)$.

Since $t \in \mathcal{T}_\ell \subseteq \mathcal{T}_{\ell-1}$, there is a node y on some shortest u-t-path P' such that $P'|_{u \to y} \in C_{\ell-1}(u)$ and thus, $y \in \overrightarrow{A}_{\ell-1}(u)$. Similarly, since $t \in \mathcal{T}_\ell$, there is a node x on some shortest y-t-path P'' such that $P''|_{y \to x} \in C_\ell(y)$ and thus, $x \in \overrightarrow{A}_\ell(y) \subseteq \mathcal{T}_\ell$. Let $P^c \in \mathcal{U}(G)$ denote the canonical shortest u-x-path. Set $P := P^c \circ P''|_{x \to t}$. Note that P is a shortest u-t-path. The definition of $\overrightarrow{A}_\ell(u)$ implies that $x \in \overrightarrow{A}_\ell(u)$. Hence, $P|_{u \to x} \in C_\ell(u)$.

By induction, this proof can be extended to multiple levels.   □

6.3.2 Computing Distance Tables

To compute an all-pairs distance table, we can use the many-to-many algorithm from Chapter 5. Roughly, this algorithm first performs independent backward searches from all transit nodes and stores the gathered distance information in *buckets* associated with each node in the search space. Then, a forward search from each transit node scans all buckets it encounters and uses the resulting path length information to update a table of tentative distances.

For the topmost table D_L (where we always have $D_L(s,t) = d(s,t)$), this procedure can be applied directly. For all other tables $D_\ell, \ell < L$, we have to respect that an explicit entry $D_\ell(s,t)$ is only required if $d(s,t) < d_{\geq\ell+1}(s,t)$—all other entries are ∞ and do not have to be explicitly stored. In order to be able to check this condition, the preprocessing of transit–node routing is done in a top–down fashion, i.e., we first compute the access nodes and the distance table for the topmost level before constructing level $L - 1$, and so on. Thus, when we compute the table D_ℓ, we can already access $d_{\geq\ell+1}(s,t)$.

A naive application of the many–to–many algorithm is prohibitive for lower levels (probably even for level $L-1$). Fortunately, there is one simple trick based on Lemma 21: when performing backward and forward searches in order to compute table D_ℓ, $\ell < L$, we do not have to relax edges out of nodes $u \in \mathcal{T}_{\ell+1}$. By this measure, we only might miss shortest s–t–paths with a node from $\mathcal{T}_{\ell+1}$ on them. However, due to Lemma 21, we already know that in these cases $d_{\geq\ell+1}(s,t) = d(s,t)$ so that $D_\ell(s,t) = \infty$.

Note that the computation of the distance table D_ℓ consists of the same local forward and backward searches as the computation of the access–node sets $\overrightarrow{A}_{\ell+1}$ and $\overleftarrow{A}_{\ell+1}$. Thus, it is sufficient to perform the respective search processes only once and extracting both the access nodes and the data required for the distance table computation.

6.3.3　Computing Locality Filters

As already mentioned, the preprocessing of transit–node routing is done in a top–down fashion. We compute the forward and backward node sets $\overrightarrow{K}_\ell$ and $\overleftarrow{K}_\ell$ first for all nodes in $\mathcal{T}_L$, then for the nodes in $\mathcal{T}_{L-1}$, and so on. For any $u \in \mathcal{T}_L$ and any level ℓ, we just have $\overrightarrow{K}_\ell(u) = \overleftarrow{K}_\ell(u) = \emptyset$. For a level $k < L$ and any node $u \in \mathcal{T}_k \setminus \mathcal{T}_{k+1}$, we 'inherit' the level–ℓ sets from the level–$(\ell - 1)$ access nodes for $\ell > k+1$ according to Equations 6.7 and 6.9; for $\ell = k+1$, we apply Equations 6.6 and 6.8. In order to deal with the latter case, we have to determine all node pairs (s,t) such that $\ell(s) = \ell(t) = k$ and $d(s,t) < d_{\geq k+1}(s,t)$. This is exactly what we do when we compute the level–k distance table D_k. Hence, the computation of the sets $\overrightarrow{K}_{k+1}$ and $\overleftarrow{K}_{k+1}$ can be viewed as a byproduct of the computation of D_k.

After all sets $\overrightarrow{K}_\ell$ and $\overleftarrow{K}_\ell$ have been determined, the locality filters are defined according to Equation 6.10.

Faster Computation of Supersets. In spite of the trick mentioned in Section 6.3.2, the computation of a distance table can get expensive so that we might want to do without distance tables in the lower levels and use some shortest-path algorithm instead that computes the required distances on demand. In this case, the locality filters can no longer be obtained as a byproduct of the distance table computation so that we have to find a different way to compute them efficiently. Let us consider some level $k < L$ and two nodes s and t such that $\ell(s) = \ell(t) = k$. Consider a local forward search from s that determines covering paths of s w.r.t. $\mathcal{T}_{k+1}$ yielding a search tree $\overrightarrow{B}$ and, analogously, a local backward search from t yielding a search tree $\overleftarrow{B}$. We set

$$\overrightarrow{K}'_{k+1}(s) := \overrightarrow{B} \setminus \mathcal{T}_{k+1} \quad \text{and} \quad \overleftarrow{K}'_{k+1}(t) := \overleftarrow{B} \setminus \mathcal{T}_{k+1}.$$

Lemma 25 $\overrightarrow{K}'_{k+1}(s) \supseteq \overrightarrow{K}_{k+1}(s)$ *and* $\overleftarrow{K}'_{k+1}(t) \supseteq \overleftarrow{K}_{k+1}(t)$.

Proof. Consider a node u from $\overrightarrow{K}_{k+1}(s)$. According to Equation 6.6, there is a node t such that u is a node on some shortest s-t-path P and $d(s,t) < d_{\geq k+1}(s,t)$. Due to Lemma 21, we can conclude that there is no shortest s-t-path with a node from $\mathcal{T}_{k+1}$ on it; in particular, $u \notin \mathcal{T}_{k+1}$. Hence, the forward search is not pruned at any node on $P|_{s \to u}$ so that $u \in \overrightarrow{B} \setminus \mathcal{T}_{k+1}$, which implies $\overrightarrow{K}'_{k+1}(s) \supseteq \overrightarrow{K}_{k+1}(s)$. An analogous proof exists for $\overleftarrow{K}'_{k+1}(t) \supseteq \overleftarrow{K}_{k+1}(t)$. $\qquad\square$

Obviously, locality filters that are based on these supersets are still correct in the sense that they fulfil Equation 6.5. However, the number of false positives increases. Note that the computation of the supersets $\overrightarrow{K}'_{k+1}(s)$ and $\overleftarrow{K}'_{k+1}(t)$ requires the same local searches as the computation of the access-node sets $\overrightarrow{A}_{k+1}(s)$ and $\overleftarrow{A}_{k+1}(t)$. Therefore, when dealing with supersets, the computation of the locality filters can be viewed as a byproduct of the computation of the access-node sets.

6.3.4 Trade-Offs

Instead of precomputing all access-node sets, distance tables, and locality filters, we can decide to compute only a part of the data required for transit-node routing and determine the remaining data on demand during the query. In case of the access nodes, we can postpone the local searches for the covering–paths set to query time. Moreover, it is sufficient to store for a node $u \in T_\ell$ only the access nodes to level $\ell + 1$; then, during a query, access nodes to higher levels can be retrieved using Equation 6.11.

In case of the distance tables, we can—as already mentioned—omit the distance tables in the lowest levels and perform an explicit shortest–path search instead.

In case of the locality filters, we can postpone the application of Equations 6.7 and 6.9 until query time as well so that a node $u \in T_\ell$ stores only $\overrightarrow{K}_{\ell+1}(u)$ and $\overleftarrow{K}_{\ell+1}(u)$.

Of course, postponing parts of the preprocessing reduces preprocessing time and memory consumption, but increases query time.

6.3.5   Outputting Complete Path Descriptions

Generally, in a graph with bounded degree (e.g., a road network) using a (near) constant time distance oracle, we can output a shortest path from s to t in (near) constant time per edge: Look for an edge (s, s') such that $d(s, s') + d(s', t) = d(s, t)$, output (s, s'). Continue by looking for a shortest path from s' to t. Repeat until t is reached.

In the special case of transit–node routing, we can speed up this process by two measures. Suppose the shortest path uses the access nodes $u \in \overrightarrow{A}_L(s)$ and $v \in \overleftarrow{A}_L(t)$. First, while reconstructing the path from s to u, we can determine the next hop by considering all adjacent nodes s' of s and checking whether $d(s, s') + d(s', u) = d(s, u)$. Usually[3], the distance $d(s', u)$ is directly available since u is also an access node of s'. Analogously, the path from v to t can be determined.

Second, reconstructing the path from u to v can work on the overlay graph G_L of G with node set T_L rather than on the original graph G. Employing the same methods that are used to expand shortcuts in case of high–way hierarchies (Section 3.4.3), we can precompute information that allows us to output the paths associated with each edge in G_L in time linear in the

---

[3]In a few cases—when u is not an access node of s' (which can only happen if the shortest paths in the graph are not unique)—, we have to consider all access nodes u' of s' and check whether $d(s, s') + d(s', u') + d(u', u) = d(s, u)$. Note that $d(u', u)$ can be looked up in the topmost distance table.

number of edges of G that it contains. Note that long distance paths will mostly consist of these precomputed paths so that the time per edge can be made very small. These techniques can be generalised to multiple levels.

6.4 A Concrete Instantiation

In this section, we give a concrete specialisation of the abstract instantiation of the previous section, determining transit–node sets using highway hierarchies (Chapter 3), performing the preprocessing using highway–node routing (Chapter 4) and the many–to–many algorithm based on highway–node routing (Chapter 5), and applying geometric circles to define the locality filters. Note that many other reasonable concrete instantiations are conceivable, which is the reason why we decided to specialise the generic algorithm from Section 6.2 in two steps instead of merging Sections 6.3 and 6.4.

6.4.1 Specifying Transit Nodes

Nodes on high levels of a highway hierarchy have the property that they are used on shortest paths far away from source and target. 'Far away' is defined with respect to the Dijkstra rank. Hence, it is natural to use (the core of) some level K of the highway hierarchy for the transit–node set T_L. Note that we have quite good (though indirect) control over the resulting size of T_L by choosing the appropriate neighbourhood sizes and the appropriate value for K. For further transit–node levels, we use (the cores of) lower levels of the highway hierarchy.

6.4.2 Computing Access Nodes

Access–node sets are computed exactly as described in Section 6.3.1 except for the fact that we use highway–node routing (based on the description in Section 4.3.4) to perform local searches in order to determine the covering–paths sets more efficiently.

This implies that before the actual preprocessing of transit–node routing is started, we have to construct a multi–level overlay graph (Sections 4.3.1 and 4.3.3) using the transit–node sets as highway–node sets.

6.4.3 Computing Distance Tables

The topmost table is determined by a standard all–pairs shortest–path com–putation (using $|\mathcal{T}_L|$–times Dijkstra's algorithm) in the topmost overlay graph G_L. Note that for the topmost level, an application of the many–to–many algorithm using the same multi–level overlay graph would be virtually equivalent to executing just $|\mathcal{T}_L|$–times Dijkstra's algorithm.

All other distance tables, however, are computed as described in Sec–tion 6.3.2, i.e., using the many–to–many algorithm from Chapter 5. At this, it is reasonable to employ an instantiation of the many–to–many algorithm that is based on the already constructed multi–level overlay graph (cp. Sec–tion 5.4).

6.4.4 Computing Locality Filters

An explicit and exact storage of the forward and backward node sets $\overrightarrow{K}_\ell$ and $\overleftarrow{K}_\ell$ would be very expensive w.r.t. memory consumption. Furthermore, we have to keep in mind that we need a very efficient operation that determines whether the intersection of two node sets is empty. For these reasons, we use *geometric circles* to represent supersets of the sets $\overrightarrow{K}_\ell$ and $\overleftarrow{K}_\ell$. We have already noted in Section 6.3.3 that using supersets of $\overrightarrow{K}_\ell$ and $\overleftarrow{K}_\ell$ still yields correct locality filters, only the number of false positives may increase.

We assume that a layout of the graph G is available, i.e., for each node in V we know its coordinates in the plane.[4] For each node u, we store forward and backward radii $\overrightarrow{\varrho}_\ell(u)$ and $\overleftarrow{\varrho}_\ell(u)$ such that

$$\overrightarrow{K}'_\ell := \{v \in V : ||v - u||_2 \leq \overrightarrow{\varrho}_\ell(u)\} \supseteq \overrightarrow{K}_\ell$$

and, analogously,

$$\overleftarrow{K}'_\ell := \{v \in V : ||v - u||_2 \leq \overleftarrow{\varrho}_\ell(u)\} \supseteq \overleftarrow{K}_\ell,$$

where $||v - u||_2$ denotes the Euclidean distance between u and v. An inter–section test can be implemented very efficiently by comparing the distance

---

[4]Even if this information is not available in the input, equally useful coordinates can be synthesised (see Section 1.2.6).

between the two involved nodes with the sum of the radii of the relevant circles:[5]

$$\overrightarrow{K}'_\ell(s) \cap \overleftarrow{K}'_\ell(t) \neq \emptyset \quad \leftrightarrow \quad ||s - t||_2 \leq \overrightarrow{\varrho}_\ell(s) + \overleftarrow{\varrho}_\ell(t). \qquad (6.12)$$

Note that the application of Equations 6.7 and 6.9 to 'inherit' node sets is quite simple using geometric circles: we use

$$\overrightarrow{\varrho}_\ell(s) := \max\{||s - u||_2 + \overrightarrow{\varrho}_\ell(u) \mid u \in \overrightarrow{A}_{\ell-1}(s)\}$$

and an analogous assignment for $\overleftarrow{\varrho}_\ell(t)$. Figure 6.5 gives an example.

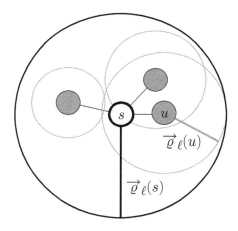

Figure 6.5: Example for the 'inheritance' of a geometric locality filter. The grey nodes constitute the set $\overrightarrow{A}_{\ell-1}(s)$.

Faster Evaluation. Combining Equations 6.10 and 6.12, we have

$$\mathcal{L}_\ell(s, t) := \bigvee_{k \leq \ell} \left(||s - t||_2 \leq \overrightarrow{\varrho}_k(s) + \overleftarrow{\varrho}_k(t) \right).$$

Thus, in order to evaluate $\mathcal{L}_\ell(s, t)$, we have to perform up to ℓ comparisons. We can easily do with only one comparison by precomputing

$$\overrightarrow{\varrho}'_\ell(s) := \max_{k \leq \ell} \overrightarrow{\varrho}_k(s) \quad \text{and} \quad \overleftarrow{\varrho}'_\ell(t) := \max_{k \leq \ell} \overleftarrow{\varrho}_k(t)$$

---

[5]To avoid the expensive square root computation that is required to determine the Euclidean distance, we can alternatively square both sides of the inequality.

and using

$$\mathcal{L}_\ell(s,t) := \left(||s - t||_2 \le \overrightarrow{\varrho}\,'_\ell(s) + \overleftarrow{\varrho}\,'_\ell(t) \right).$$

Note that the number of false positives may increase.

6.4.5   Hasty Inheritance

In order to accelerate the preprocessing, we have already made extensive use of the idea of handing down obtained data (access nodes, locality filters) to lower levels. Basically, for a node u in a level ℓ, we look for covering paths w.r.t. $\mathcal{T}_{\ell+1}$ and inherit the data stored at the endpoints of the covering paths.

We can think of a hastier approach: When we search for the covering paths of u and encounter a node v that has already been processed, i.e., that has already adopted the data from level $\ell + 1$, we do not have to continue the search from v. Instead, we can directly inherit the data from v.

In our implementation, we use this technique when we hand data down from level 1 to level 0.

6.4.6   An Economical and a Generous Variant

In our experiments, we consider two different variants as illustrated in Figure 6.6.

Variant 'Economical'   aims at a good compromise between space consumption, preprocessing time and query time. It uses three levels on top of the original graph (i.e., $L = 3$). We make extensively use of the options presented in Section 6.3.4. At each node $u \in \mathcal{T}_2$, we store the access nodes to level 3, and at each node $u \in V$, we store the access nodes to level 2. This means that the level-3 access nodes for nodes $u \notin \mathcal{T}_2$ have to be reconstructed during query using Equation 6.11. Level 1 is only used to accelerate the preprocessing (since it is faster to compute access nodes and locality filters only for a subset $\mathcal{T}_1 \subseteq V$, handing the data down to all nodes). We do not use level-1 access nodes or a level-1 distance table. Instead, we just set $\mathcal{L}_1(s,t) :=$ true for all node pairs (s,t) so that if the query reaches level 1, it is automatically forwarded to level 0.

We explicitly store the level-2 and level-3 distance tables. In level 0, instead of keeping a distance table, we perform a shortest-path query using highway-node routing.

The locality filters are dealt with analogously to the access nodes: at each node $u \in \mathcal{T}_2$, we store $\overrightarrow{\varrho}_3(u)$ and $\overleftarrow{\varrho}_3(u)$, and at each node $u \in V$, we store $\overrightarrow{\varrho}'_2(u)$ and $\overleftarrow{\varrho}'_2(u)$. The level-2 locality filter is determined with the fast but less precise method described at the end of Section 6.3.3.

Variant 'Generous'   is tuned for very fast query times. As the economical variant, it uses three levels on top of the original graph (but in this case, level 1 is not only used to accelerate the preprocessing). At each node u, we store the access nodes and the locality filters[6] required for the query in level 2 and 3. This allows direct access to these levels. For level 1, we store neither access nodes nor a locality filter. Instead, if required, we perform local searches to determine the access nodes and we use the trivial locality filter $\mathcal{L}_1(s,t) :=$ true for all node pairs (s,t). We explicitly store the level–1–3 distance tables, while we perform a shortest-path query in level 0 (if required). Note that having a level-1 distance table is a significant difference from the economical variant. Interestingly, the search for the level-1 access nodes already involves the search in level 0 so that no extra work is imposed by the level-0 search. This also explains why it is reasonable to just set $\mathcal{L}_1(s,t) :=$ true.

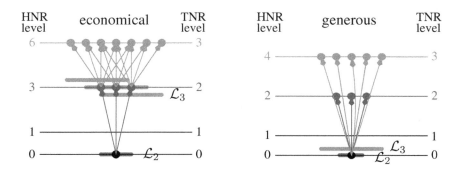

Figure 6.6: Two variants of transit-node routing (TNR) based on highway-node routing (HNR).

---

[6]i.e., the radii $\overrightarrow{\varrho}'_2(u)$, $\overleftarrow{\varrho}'_2(u)$, $\overrightarrow{\varrho}'_3(u)$, and $\overleftarrow{\varrho}'_3(u)$

6.4.7   Queries

Queries are performed in a top–down fashion. For a given query pair (s,t), first $\overrightarrow{A}_3(s)$ and $\overleftarrow{A}_3(t)$ are either looked up or computed depending on the used variant. Then table lookups in the top–level distance table yield a first guess for $d(s,t)$. Now, if $\neg\mathcal{L}_3(s,t)$, we are done. Otherwise, the same procedure is repeated for level 2. If even $\mathcal{L}_2(s,t)$ is true, we perform a bidirectional shortest–path search using highway–node routing that can stop if both the forward and backward search radius (i.e., the key of the minimum element in the respective priority queue) exceed the upper bound computed in levels 2 and 3. Furthermore, the search need not expand from any node $u \in \mathcal{T}_2$ since paths going over these nodes are covered by the search in levels 2 and 3. In the generous variant, the search is already stopped at the level–1 access nodes, and additional lookups in the level–1 distance table are performed.

6.4.8   Outputting Complete Path Descriptions

The general methods from Section 6.3.5 can be applied rather directly to our concrete instantiation in order to determine a complete description of the shortest path. To unpack shortcuts, we can fall back on the routines used in the highway hierarchies approach (Section 3.4.3).

6.5   Concluding Remarks

Review. Transit–node routing provides the fastest available query times for large static real–world road networks. Speedups compared to Dijkstra's algorithm exceed factor one million. The extremely good query performance does not imply prohibitive preprocessing times or memory consumption. In fact, the preprocessing is still clearly faster than many other route planning techniques that achieve considerably smaller speedups. Moreover, transit–node routing is not only optimised for long–range queries, but also answers local queries very efficiently.

Alternative Instantiations. There seem to be two basic approaches to transit–node routing. One that starts with a locality filter $\mathcal{L}$ and then has to find a good set of transit nodes $\mathcal{T}$ for which $\mathcal{L}$ works (e.g., the grid-based implementation [3]). The other approach starts with $\mathcal{T}$ and then has to find a locality filter that can be efficiently evaluated and detects as accurately as possible whether local search is needed (e.g., our abstract and concrete instantiations, Sections 6.3–6.4). Both basic approaches fit in the generic framework introduced in Section 6.2. In [4, 71], we describe a few additional general preprocessing techniques that might be useful for instantiations that differ from the one specified in Section 6.3.

Future Work. Like in the case of highway-node routing, we expect that it might be possible to determine even better transit-node sets. Ideally, this could imply an improvement w.r.t. preprocessing time, memory consumption, and query times.

Locality Filters. The observation that in the past, successful geometric speedup techniques have always been beaten by related non-geometric techniques (e.g. geometric A^* search by landmark-based A^* search or geometric containers by edge flags) suggests the hypothesis that geometry is not required for efficient route planning algorithms. Against this background, the fact that we have to use a *geometric* locality filter to obtain our best results is dissatisfying[7] since it *contradicts* our hypothesis. Contrariwise, if we could do without a geometric filter, our transit-node routing instantiation would *confirm* our hypothesis since its query times are superior to those of the implementation that is based on a grid division and thus, on geometry.

An alternative, non-geometric locality filter could exploit the fact that nodes that are so close that a local query is required *usually* share common access nodes. Such a filter could be evaluated very efficiently since it is equivalent to checking whether some lookup in the distance table returns a zero.[8] Preliminary experiments indicate that this locality filter would yield less false positives. However, the remaining difficulty is hidden in the word 'usually': there are a few exceptions where a local node pair does not share a common access node. In order to get a correct locality filter, we have to deal with these exceptions appropriately, perhaps by storing and checking

---

[7]This is *not* a practical problem, but it is undesirable *as a matter of principle.*
[8]At least if we disallow zero weight edges, getting a zero implies that forward and backward access nodes match.

them explicitly[9] or by deliberately adding 'superfluous' access nodes in the few exceptional cases to ensure the 'common–access–node rule'. Correctly handing down access nodes and locality filters to lower levels makes an implementation of this idea nontrivial.

Reducing the Number of Table Lookups.   For a given source–target pair (s,t), let

$$a := \max(|\overrightarrow{A}_L(s)|, |\overleftarrow{A}_L(t)|).$$

For a global query (i.e., $\mathcal{L}_L(s,t) = \mathsf{false}$), we need $O(a)$ time to lookup all access nodes, $O(a^2)$ to perform the table lookups, and $O(1)$ to check the locality filter. Thus, if we want to further improve the query times, the first attempt should be to reduce the number of table lookups. This could be done by excluding certain access nodes at the outset, using an idea very similar to the edge flag approach (Section 1.2.2). We partition the topmost overlay graph G_L into k regions and store for each node $u \in \mathcal{T}_L$ its region $r(u) \in \{1, \ldots, k\}$. For each node s and each access node $u \in \overrightarrow{A}_L(s)$, we manage a flag vector $f_{s,u} : \{1, \ldots, k\} \rightarrow \{\mathsf{true}, \mathsf{false}\}$ such that $f_{s,u}(x)$ is true iff there is a shortest path from s via u to some node $v \in \mathcal{T}_L$ with $r(v) = x$. These flag vectors can be precomputed in the following way, again using ideas similar to those used in the preprocessing of the edge flag approach: Let $B \subseteq \mathcal{T}_L$ denote the set of border nodes, i.e., nodes that are adjacent to some node in G_L that belongs to a different region. For each node $s \in V$ and each border node $b \in B$, we determine the access node $u \in \overrightarrow{A}_L(s)$ that minimises $d(s,u) + d(u,b)$; we set $f_{s,u}(r(b))$ to true. In addition, $f_{s,u}(r(u))$ is set to true for each $s \in V$ and each access node $u \in \overrightarrow{A}_L(s)$ since each access node obviously minimises the distance to itself. We assume that the preprocessing can be done sufficiently fast since $|\mathcal{T}_L|$ is already small, $|B|$ is even smaller, and the distances $d(u,b)$ can be looked up in the topmost distance table. An analogous preprocessing step has to be done for the backward direction. Presumably, it is a good idea to just store the bitwise OR of the forward and backward flag vectors in order to keep the memory consumption within reasonable bounds.

Then, in a query from s to t, we can take advantage of the precomputed flag vectors. First, we consider all backward access nodes of t and build

---

[9]Note that *identifying* all exceptions during preprocessing is rather simple.

the flag vector f_t such that $f_t(r(u)) = \text{true}$ for each $u \in \overleftarrow{A}_L(t)$. Second, we consider only forward access nodes u of s with the property that the bit-wise AND of $f_{s,u}$ and f_t is not zero; we denote this set by $\overrightarrow{A}'_L(s)$; during this step, we also build the vector f_s such that $f_s(r(u)) = \text{true}$ for each $u \in \overrightarrow{A}'_L(s)$. Third, we use f_s to determine the subset $\overleftarrow{A}'_L(t) \subseteq \overleftarrow{A}_L(t)$ analogously to the second step. Now, it is sufficient to perform only $|\overrightarrow{A}'_L(s)| \times |\overleftarrow{A}'_L(t)|$ table lookups. We conjecture—based on the excellent sense of goal direction that the edge flag approach exhibits—that by this means, the number of table lookups can be reduced from around 75 to 1–4. Note that determining $\overrightarrow{A}'_L(s)$ and $\overleftarrow{A}'_L(t)$ is in $O(a)$, in particular operations on the flag vectors can be considered as quite cheap.

The preprocessing of the flag vectors can be further accelerated: First, we can perform the computations only from nodes $s \in \mathcal{T}_1$ (instead of considering all nodes) and hand the obtained flag vectors down in an appropriate way. However, that way, the effectiveness of the flag vectors could be impaired. Second, we could rearrange the columns of the distance table so that all border nodes are stored consecutively, which should reduce the number of cache misses during preprocessing.

References. This chapter is partly based on [71, 4, 5]. The relation between our contributions and the work by Bast, Funke, and Matijevic, who introduced transit–node routing in the context of a grid–based implementation in [3] (which was followed by two joint publications [4, 5]), has been explained in Sections 1.2.3 and 1.3.5.

In this thesis, the formal representation of the generic framework has been largely extended compared to earlier publications. Furthermore, we now have both an implementation based on highway hierarchies and a more recent one based on highway–node routing. Since the new implementation is superior to the old one (in particular w.r.t. preprocessing times) and more flexible (in particular w.r.t. choosing the transit–node sets), we concentrated on the new one in this thesis. Details on the highway–hierarchy–based instantiation including experimental results can be found in [71, 4, 5].

7

Experiments

7.1 Implementation

We have implemented all of our route planning techniques closely following the specifications from Chapters 3–6. The implementation has been done in C++ using basic data structures from the C++ Standard Template Library. Since our various approaches are closely related, there is a high potential of sharing common code. Therefore, we decided to write a single program that unites the functionality of all route planning techniques. In order to avoid runtime overheads, we make extensive use of *generic programming* techniques using C++'s template class mechanism. This allows, for example, to represent several variants of Dijkstra's algorithm (which most of our preprocessing and query algorithms are based on) in a single template class without losing performance in comparison to having each variant in a separate class.[1] The obvious advantage of shared code is that improving/tuning one variant often also yields an immediate improvement of the other variants. Overall, our program consists of more than 20 000 lines of code (including comments).

We put particular efforts into carefully implementing efficient data structures, e.g., for representing graphs. We decided against using existing libraries like LEDA [58] or the Boost Graph Library [82] since the generality

---

[1] Actually, in the current version there are no less than 36 different variants.

of such libraries entail certain undesired overheads.[2]

To obtain a robust implementation, we include extensive consistency checks in assertions and perform experiments that are checked against reference implementations, i.e., queries are checked against Dijkstra's algorithm and fast preprocessing algorithms are checked against naive implementations. Moreover, we created our own visualisation tools [9] that can handle large graphs and are able to illustrate our route planning approaches. By this means, several possibilities for further improvements have been discovered and utilised.

We use 32-bit integers to store edge weights and path lengths. For more details on the implementation, in particular on the employed data structures, we refer to Appendix A.

7.2 Experimental Setting

7.2.1 Environment

The experiments were done on one core of a single AMD Opteron Processor 270 clocked at 2.0 GHz with 8 GB main memory and 2×1 MB L2 cache, running SuSE Linux 10.0 (kernel 2.6.13). The program was compiled by the GNU C++ compiler 4.0.2 using optimisation level 3. Results for the DIMACS Challenge benchmarks [1] can be found in Table 7.1.

7.2.2 Instances

Main Instances. We deal with the road networks of Western Europe[3] (which we often just call '*Europe*') and of the USA (without Hawaii) and Canada (*USA/CAN*). Both networks have been made available for scientific use by the company PTV AG. For each edge, its length and one out

---

[2]Furthermore, in the past, various problems have been reported when using graph libraries and dealing with very large instances. In the meantime, it seems that the situation has improved. For example, the current implementation of the edge flag approach [36] uses the Boost Graph Library and is able to handle very large road networks. However, some runtime overheads still remain.

[3]Austria, Belgium, Denmark, France, Germany, Italy, Luxembourg, the Netherlands, Norway, Portugal, Spain, Sweden, Switzerland, and the UK

Table 7.1: DIMACS Challenge benchmarks [1] for US (sub)graphs (query time [ms]).

graph	metric	
	time	dist
NY	29.6	28.5
BAY	34.7	33.3
COL	51.5	49.0
FLA	134.8	120.5
NW	161.1	146.1
NE	225.4	197.2
CAL	291.1	235.4
LKS	461.3	366.1
E	681.8	536.4
W	1 211.2	988.2
CTR	4 485.7	3 708.1
USA	5 355.6	4 509.1

of 13 road categories (e.g., motorway, national road, regional road, urban street) is provided.

In addition, we perform experiments on a publicly available version of the US road network (without Alaska and Hawaii) that was obtained from the TIGER/Line Files [92] (*USA (Tiger)*). However, in contrast to the PTV data, the TIGER graph is undirected, planarised and distinguishes only between four road categories, in fact 91% of all roads belong to the slowest category so that you cannot discriminate them.

Strongly Connected Components.   For the 9th DIMACS Implementation Challenge [1], our road networks of Europe and the USA (Tiger) were established as benchmark instances. However, since not all participants could handle unconnected graphs, in each case, only the largest strongly connected component was considered. Although our implementation is able to handle unconnected graphs, we restricted ourselves to the strongly connected components in case of the combination of highway hierarchies with goal-directed search (Section 7.5) and in case of transit-node routing (Section 7.9) to comply with the guidelines of the challenge. In both cases, it makes hardly any difference[4] which version of the graph is used since the largest strongly connected component consists of about 99% of all nodes.

Different Metrics. For most practical applications, a *travel time* metric is most useful, i.e., the edge weights correspond to an estimate of the travel time that is needed to traverse the edge. In order to compute the edge weights, we assign an average speed to each road category (see Table 7.2).

Table 7.2: Average Speeds [km/h]. The last column contains the average speed for "forest roads, pedestrian zones, private roads, gravel roads or other roads not suitable for general traffic".

	motorway			national			regional			urban			
	fast		slow	fast		slow	fast		slow	fast		slow	
Europe	130	120	110	100	90	80	70	60	50	40	30	20	10
USA/CAN	112	104	96	96	88	80	72	64	56	40	32	24	16
USA (Tiger)		100			80			60			40		

In some cases, we also deal with a *distance* metric (where we directly use the provided lengths) and a *unit* metric (where each edge gets weight 1).

An Even Larger Road Network. Very recently, we obtained a new version of the European road network (*New Europe*) that is larger than the old one and covers more countries[5]. It has been provided for scientific use by the company ORTEC. So far, we have done only a few experiments on it. Unless otherwise stated, the term '(Western) Europe' always refers to the smaller network provided by PTV.

Table 7.3 gives the sizes of the used road networks.

---

[4]Note that in case of transit-node routing, we could get a small improvement w.r.t. query times if we changed our implementation such that only connected inputs are handled correctly—because in this case, we could omit a few checks for the special case that two nodes are not connected.

[5]In addition to the old version, the Czech Republic, Finland, Hungary, Ireland, Poland, and Slovakia.

Table 7.3: Test Instances. In case of Europe and the USA (Tiger), we give the size of both variants: the original one and the largest strongly connected component (scc).

road network	#nodes	#directed edges
Europe	18 029 721	42 199 587
Europe (scc)	18 010 173	42 188 664
USA/CAN	18 741 705	47 244 849
USA (Tiger)	24 278 285	58 213 192
USA (Tiger) (scc)	23 947 347	57 708 624
New Europe	33 726 989	75 108 089

7.2.3 Preliminary Remarks

Unless otherwise stated, the experimental results refer to the scenario where the *travel time* metric is used, only the shortest-path *length* is computed without outputting the actual route, and turning restrictions are *ignored*.

When we specify the memory consumption of one of our approaches, we usually give the *overhead*, which accounts for the *additional* memory that is needed by our approach compared to a space-efficient bidirectional implementation of Dijkstra's algorithm. This overhead is always expressed in 'bytes per node'. Alternatively, we sometimes give the *total disk space* (in MB), which is the space that is needed to store the original graph together with the preprocessed auxiliary data on hard disk. It does not include volatile data structures like the priority queues.

7.3 Methodology

7.3.1 Random Queries

As a simple, widely used and accepted performance measure, we run queries using source-target pairs that are picked *uniformly at random*. The advantage of this measure is that it can be expressed by a single figure (the average query time) and that it is independent of a particular application. In addition to the average query time, we also often given the average search space size and the average number of relaxed edges. As basis for comparisons, we use Dijkstra's algorithm: the term '*speedup*' refers to the ratio between the average query time or the average search space size of Dijkstra's algorithm and the corresponding measurement of the algorithm whose performance is studied. Unless otherwise stated, in our experiments, we pick $10\,000$ random source–target pairs.

7.3.2 Local Queries

For use in applications it is unrealistic to assume a uniform distribution of queries in large graphs such as Europe or the USA. On the other hand, it would be hardly more realistic to arbitrarily cut the graph into smaller pieces. Therefore, we decided to also measure *local queries* within the big graphs: We choose random sample points s and for each power of two $r = 2^k$, we use Dijkstra's algorithm to find the node t with Dijkstra rank $\mathrm{rk}_s(t) = r$. We then use our algorithm to make an s–t–query. By plotting the resulting statistics for each value $r = 2^k$, we can see how the performance scales with a natural measure of difficulty of the query. We represent the distributions as box-and-whisker plots [67]: each box spreads from the lower to the upper quartile and contains the median, the whiskers extend to the minimum and maximum value omitting outliers, which are plotted individually. Such plots are based on $1\,000$ random sample points s. For some examples, see Figures 7.3, 7.6, 7.8, and 7.12.

7.3.3 Worst Case Upper Bounds

For any bidirectional approach where forward and backward search can be executed independently of each other—this applies both to highway hierarchies and highway-node routing—, we can use the following technique to obtain a *per-instance worst-case guarantee*, i.e., an upper bound on the search space size for any possible point-to-point query for a given fixed graph G: By executing a query from each node of G to an added isolated dummy node and a query from the dummy node to each actual node in the backward graph, we obtain a distribution of the search space sizes of the forward and backward search, respectively. We can combine both distributions to get an upper bound for the distribution of the search space sizes of bidirectional queries: when $\mathcal{F}_{\rightarrow}(x)$ $(\mathcal{F}_{\leftarrow}(x))$ denotes the number of source (target) nodes whose search space consists of x nodes in a forward (backward) search, we define

$$\mathcal{F}_{\leftrightarrow}(z) := \sum_{x+y=z} \mathcal{F}_{\rightarrow}(x) \cdot \mathcal{F}_{\leftarrow}(y), \tag{7.1}$$

i.e., $\mathcal{F}_{\leftrightarrow}(z)$ is the number of s-t-pairs such that the upper bound of the search space size of a query from s to t is z. In particular, we obtain the upper bound $\max\{z \mid \mathcal{F}_{\leftrightarrow}(z) > 0\}$ for the worst case without performing all n^2 possible queries. Examples can be found in Figures 7.4 and 7.9.

For bidirectional approaches that employ a distance table—this ap-plies both to highway hierarchies and transit-node routing—, a very similar method can be used to derive histograms for the number of accessed table entries: Let $\mathcal{F}'_{\rightarrow}(x)$ denote the number of source nodes with x forward en-trance points to the topmost core (in case of highway hierarchies) or with x forward level-L access nodes (in case of transit-node routing). $\mathcal{F}'_{\leftarrow}(x)$ is defined analogously. Then, a distribution for the number of table accesses can be derived by replacing $x + y = z$ with $x \cdot y = z$ in Equation 7.1 and thus, using

$$\mathcal{F}'_{\leftrightarrow}(z) := \sum_{x \cdot y = z} \mathcal{F}'_{\rightarrow}(x) \cdot \mathcal{F}'_{\leftarrow}(y). \tag{7.2}$$

Figures 7.5 and 7.13 have been obtained by this means.

7.4 Highway Hierarchies

7.4.1 Parameters

Default Settings. Unless otherwise stated, the following default settings apply. We use the *maverick factor* $f = 2(i - 1)$ for the i-th iteration of the construction procedure, the contraction rate $c = 2$, the shortcut hops limit 10, and the neighbourhood size $H = 30$ for Europe and $H = 40$ for both North American networks—the same neighbourhood size is used for all levels of a hierarchy. First, we contract the original graph.[6] Then, we perform five iterations of our construction procedure, which determines a highway network and its core. Finally, we compute the distance table for all level-5 core nodes.

---

[6]In Section 3.2, we gave the definition of the highway hierarchies where we first construct a highway network and then contract it. We decided to change this order in the experiments, i.e., to start with an initial contraction phase, since we observed a better performance in this case.

Self-Similarity.   For two levels ℓ and $\ell + 1$ of a highway hierarchy, the *shrinking factor* is the ratio between $|E'_\ell|$ and $|E'_{\ell+1}|$. In our experiments, we observed that the highway hierarchies of Europe and the USA were almost *self-similar* in the sense that the shrinking factor remained nearly unchanged from level to level when we used the same neighbourhood size H for all levels—provided that H was not too small.

Figure 7.1 demonstrates the shrinking process for Europe. Note that the first contraction step is not shown. In contrast to our default settings, we do not stop after five iterations. For most levels and $H \geq 70$, we observe an almost constant shrinking factor[7] (which appears as a straight line due to the logarithmic scale of the y-axis). The greater the neighbourhood size, the greater the shrinking factor. The last iteration is an exception: the highway network collapses, i.e., it shrinks very fast because nodes that are close to the border of the network usually do not belong to the next level of the highway hierarchy, and when the network gets small, almost all nodes are close to the border. In case of the smallest neighbourhood size ($H = 30$), the shrinking factor gets so small that the network does not collapse even after 14 levels have been constructed.

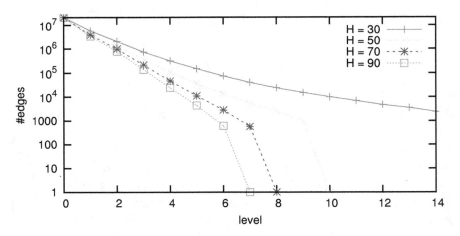

Figure 7.1: Shrinking of the highway networks of Europe. For different neighbourhood sizes H and for each level ℓ, we plot $|E'_\ell|$, i.e., the number of edges that belong to the core of level ℓ.

---

[7]Detailed numbers for $H = 70$ can be found in Table 3.1 in Section 3.1.

Varying the Neighbourhood Size. Note that in order to simplify the experimental setup, all results in the remainder of Section 7.4.1 have been obtained without rearranging nodes by level. This simplification is unproblematic since we want to demonstrate the effects of choosing different parameter settings and at this, the *relative* performance is already very meaningful.

In one test series (Figure 7.2), we used all the default settings except for the neighbourhood size H, which we varied in steps of 5. On the one hand, if H is too small, the shrinking of the highway networks is less effective so that the level–5 core is still quite big. Hence, we need much time and space to precompute and store the distance table. On the other hand, if H gets bigger, the time needed to preprocess the lower levels increases because the area covered by the local searches depends on the neighbourhood size. Furthermore, during a query, it takes longer to leave the lower levels in order to get to the topmost level where the distance table can be used. Thus, the query time increases as well. We observe that the preprocessing time is minimised for neighbourhood sizes around 40. In particular, the optimal neighbourhood size does not vary very much from graph to graph. In other words, if we used the same parameter H, say 40, for all road networks, the resulting performance would be very close to the optimum. Obviously, choosing different neighbourhood sizes leads to different space–time trade–offs.

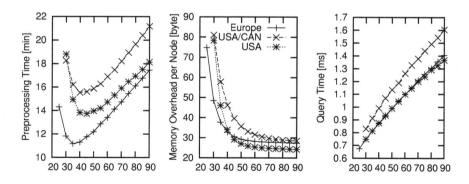

Figure 7.2: Preprocessing and query performance depending on the neighbourhood size H.

Varying the Contraction Rate. In another test series (Table 7.4), we did not use a distance table, but repeated the construction process until the top–most level was empty or the hierarchy consisted of 15 levels. We varied the contraction rate c from 0.5 to 2.5. In case of $c = 0.5$ (and $H = 30$), the shrinking of the highway networks does not work properly so that the topmost level is still very big. This yields huge query times. Choosing larger contraction rates reduces the preprocessing and query times since the cores and search spaces get smaller. However, the memory usage and the average degree are slightly increased since more shortcuts are introduced. Adding too many shortcuts ($c = 2.5$) further reduces the search space, but the number of relaxed edges increases so that the query times get worse.

Table 7.4: Preprocessing and query performance for the European road net–work depending on the contraction rate c. 'overhead' denotes the average memory overhead per node in bytes.

contr. rate c	PREPROCESSING			QUERY		
	time [min]	over–head	∅deg.	time [ms]	#settled nodes	#relaxed edges
0.5	83	30	3.2	391.73	472 326	1 023 944
1.0	15	28	3.7	5.48	6 396	23 612
1.5	11	28	3.8	1.93	1 830	9 281
2.0	11	29	4.0	1.85	1 542	8 913
2.5	11	30	4.1	1.96	1 489	9 175

Varying the Number of Levels. In a third test series (Table 7.5), we used the default settings except for the number of levels, which we varied from 6 to 11. Note that the original graph and its core (i.e., the result of the first contraction step) counts as one level so that for example '6 levels' means that only five levels are constructed. In each test case, a distance table was used in the topmost level. The construction of the higher levels of the hier–archy is very fast and has no significant effect on the preprocessing times. In contrast, using only six levels yields a rather large distance table, which somewhat slows down the preprocessing and increases the memory usage. However, in terms of query times, '6 levels' is the optimal choice since us–ing the distance table is faster than continuing the search in higher levels. We omitted experiments with less levels since this would yield very large distance tables consuming very much memory.

Table 7.5: Preprocessing and query performance for the European road network depending on the number of levels. 'overhead' denotes the average memory overhead per node in bytes.

# levels	PREPROC. time [min]	over– head	QUERY time [ms]	#settled nodes
6	12	48	0.75	709
7	10	34	0.93	852
8	10	30	1.14	991
9	10	30	1.35	1 123
10	10	29	1.54	1 241
11	10	29	1.67	1 326

Results for further combinations of neighbourhood size, contraction rate, and number of levels can be found in Tables 7.7 and 7.8.

7.4.2 Main Results

Table 7.6 summarises the key results of the experiments where we apply the default parameters and perform random queries (as specified in Section 7.3.1).

Table 7.6: Overview of the key results. Note that 'worst case' is an upper bound for *any* possible query in the respective graph and *not* only within the chosen sample (cp. Section 7.4.5).

		Europe	USA/CAN	USA (Tiger)
PARAM.	H	30	40	40
PREPROC.	CPU time [min]	13	17	15
	∅overhead/node [byte]	48	46	34
QUERY	CPU time [ms]	0.61	0.83	0.67
	#settled nodes	709	871	925
	#relaxed edges	2 531	3 376	3 823
	speedup (CPU time)	9 935	7 259	9 303
	speedup (#settled nodes)	12 715	10 750	12 889
	worst case (#settled nodes)	2 388	2 428	2 505

Table 7.7: Preprocessing and query performance for the European road net‐
work depending on the contraction rate c and the neighbourhood size H.
We do not use a distance table, but repeat the construction process until the
topmost level is empty or the hierarchy consists of 15 levels.

contr.	nbh.	PREPROCESSING			QUERY		
rate c	size H	time [min]	over‐head	∅deg.	time [ms]	#settled nodes	#relaxed edges
	30	83	30	3.2	391.73	472 326	1 023 944
	40	83	28	3.2	267.57	334 287	711 082
	50	87	27	3.2	188.55	242 787	506 543
0.5	60	86	27	3.2	135.27	177 558	362 748
	70	87	26	3.2	101.36	135 560	271 324
	80	89	26	3.1	73.40	99 857	196 150
	90	87	25	3.1	55.02	75 969	146 247
	30	15	28	3.7	5.48	6 396	23 612
	40	15	28	3.7	2.62	3 033	11 315
	50	17	27	3.6	2.13	2 406	8 902
1.0	60	18	27	3.6	1.93	2 201	8 001
	70	19	26	3.6	1.80	2 151	7 474
	80	20	26	3.6	1.79	2 193	7 392
	90	22	26	3.6	1.78	2 221	7 268
	30	11	28	3.8	1.93	1 830	9 281
	40	12	28	3.8	1.72	1 628	7 672
	50	13	27	3.7	1.56	1 593	6 975
1.5	60	14	27	3.7	1.53	1 645	6 697
	70	15	27	3.7	1.51	1 673	6 590
	80	17	27	3.7	1.51	1 726	6 719
	90	18	27	3.7	1.54	1 782	6 655
	30	11	29	4.0	1.85	1 542	8 913
	40	11	29	3.9	1.64	1 475	7 646
	50	12	28	3.9	1.48	1 470	6 785
2.0	60	14	28	3.8	1.46	1 506	6 650
	70	15	28	3.8	1.45	1 547	6 649
	80	16	27	3.8	1.49	1 611	6 935
	90	17	27	3.8	1.53	1 675	6 988
	30	11	30	4.1	1.96	1 489	9 175
	40	11	29	4.0	1.70	1 453	7 822
	50	12	29	4.0	1.58	1 467	7 119
2.5	60	14	29	3.9	1.57	1 493	7 035
	70	15	28	3.9	1.54	1 536	6 905
	80	16	28	3.9	1.55	1 583	7 094
	90	18	28	3.9	1.58	1 645	7 204

Table 7.8: Preprocessing and query performance for the European road net-work depending on the number of levels and the neighbourhood size H. In the topmost level, a distance table is used.

#levels	nbh. size H	PREPROCESSING time [min]	PREPROCESSING over-head	PREPROCESSING ∅deg.	QUERY time [ms]	QUERY #settled nodes	QUERY #relaxed edges
	40	14	60	3.9	0.67	691	2 398
	50	13	40	3.9	0.77	818	2 892
5	60	14	32	3.8	0.87	938	3 361
	70	15	30	3.8	0.96	1 058	3 837
	80	16	28	3.8	1.05	1 165	4 278
	90	17	28	3.8	1.13	1 269	4 697
	30	12	48	4.0	0.75	709	2 531
	40	11	33	3.9	0.87	867	3 171
	50	12	29	3.9	0.99	1 015	3 759
6	60	13	28	3.8	1.10	1 157	4 299
	70	15	28	3.8	1.21	1 292	4 837
	80	16	28	3.8	1.30	1 414	5 311
	90	17	27	3.8	1.40	1 521	5 817
	30	10	34	4.0	0.93	852	3 195
	40	11	29	3.9	1.07	1 025	3 894
	50	12	28	3.9	1.20	1 187	4 538
7	60	13	28	3.8	1.32	1 344	5 166
	70	15	28	3.8	1.39	1 462	5 689
	80	16	27	3.8	1.47	1 578	6 179
	90	18	27	3.8	1.53	1 668	6 661
	30	10	30	4.0	1.14	991	3 853
	40	11	29	3.9	1.27	1 171	4 624
	50	12	28	3.9	1.36	1 321	5 283
8	60	14	28	3.8	1.43	1 455	5 887
	70	15	28	3.8	1.46	1 546	6 338
	80	16	27	3.8	1.48	1 611	6 935
	90	18	27	3.8	1.53	1 675	6 988
	30	10	30	4.0	1.35	1 123	4 532
	40	11	29	3.9	1.45	1 289	5 338
9	50	12	28	3.9	1.48	1 417	5 931
	60	14	28	3.8	1.47	1 506	6 429
	70	15	28	3.8	1.46	1 547	6 649
	30	10	29	4.0	1.54	1 241	5 214
10	40	11	29	3.9	1.57	1 380	6 012
	50	12	28	3.9	1.51	1 468	6 470
	60	14	28	3.8	1.46	1 506	6 650

7.4.3 Local Queries

Figure 7.3 shows the query times according to the methodology introduced in Section 7.3.2. Note that for ranks up to 2^{18} the median query times are scaling quite smoothly and the growth is much slower than the exponential increase we would expect in a plot with logarithmic x axis, linear y axis, and any growth rate of the form r^ρ for Dijkstra rank r and some constant power ρ; the curve is also not the straight line one would expect from a query time logarithmic in r. For ranks $r \geq 2^{19}$, the query times hardly rise due to the fact that the all-pairs distance table can bridge the gap between the forward and backward search of these queries irrespective of dealing with a small or a large gap. In case of Europe and USA/CAN, the query times drop for $r = 2^{24}$ since r is only slightly smaller than the number of nodes so that the target lies close to the border of the respective road network which implies some kind of trivial sense of goal direction for the backward search (because, in the beginning, we practically cannot go into the wrong direction).

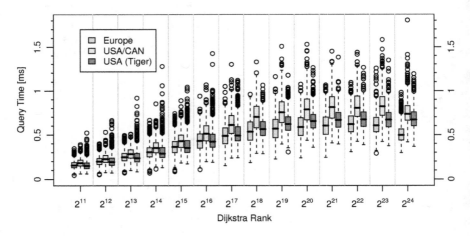

Figure 7.3: Local queries.

7.4.4 Space Saving

If we omit the first contraction step and use a smaller contraction rate ($\Rightarrow$ less shortcuts), use a bigger neighbourhood size ($\Rightarrow$ higher levels get smaller), and construct more levels before the distance table is used

($\Rightarrow$ smaller distance table), the memory usage can be reduced considerably. In case of Europe, using seven levels, $H = 100$, and $c = 1$ yields an average overhead per node of 17 bytes. The construction and query times increase to 55 min and 1.10 ms, respectively.

7.4.5 Worst Case Upper Bounds

We apply the techniques introduced in Section 7.3.3. Figure 7.4 visualises the distribution of the search space sizes as a histogram.

Similarly, Figure 7.5 represents the distribution of the number of entries in the distance table that have to be accessed during an s–t–query. While the average values are reasonably small (4 066 in case of Europe), the worst case can get quite large (62 379). For example, accessing 62 379 entries in a table of size 9 351 × 9 351 takes about 1.1 ms, where 9 351 is the number of nodes of the level-5 core of the European highway hierarchy. Hence, in some cases the time needed to determine the optimal entry in the distance table might dominate the query time. We could try to improve the worst case by introducing a case distinction that checks whether the number of entries that have to be considered exceeds a certain threshold. If so, we would not use the distance table, but continue with the normal search process. However, this measures would have only little effect on the *average* performance.

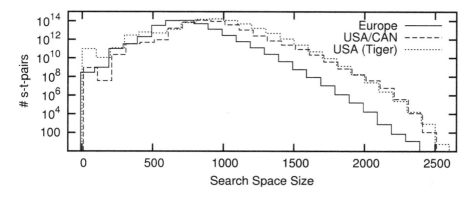

Figure 7.4: Histogram of upper bounds for the search space sizes of s–t– queries. To increase readability, only the outline of the histogram is plotted instead of the complete boxes.

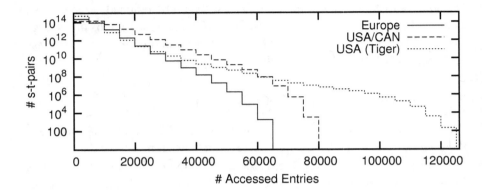

Figure 7.5: Histogram of upper bounds for the number of entries in the distance table that have to be accessed during an s-t-query.

7.4.6 Outputting Complete Path Descriptions

So far, we have reported only the times needed to compute the shortest-path length between two nodes. Now, we determine a complete description of the shortest path. In Table 7.9 we give the additional preprocessing time and the additional disk space for the unpacking data structures. Furthermore, we report the additional time that is needed to determine a complete description of the shortest path and to traverse[8] it summing up the weights of all edges as a sanity check—assuming that the query to determine the shortest-path length has already been performed. That means that the total average time to determine a shortest path is the time given in Table 7.9 plus the query time given in previous tables[9]. Note that Variant 1 is no longer supported by the current version of our implementation so that the numbers in the first data row of Table 7.9 have been obtained with an older version and different settings.

We can conclude that even Variant 3 requires little additional preprocessing time and only a moderate amount of space. With Variant 3, the time for outputting the path remains considerably smaller than the time to deter-

---

[8]Note that we do *not* traverse the path in the original graph, but we directly scan the assembled description of the path.

[9]Note that in the current implementation outputting path descriptions and the feature to rearrange nodes by level are mutually exclusive. However, this is not a limitation in principle.

mine the path length and a factor 3–5 smaller than using Variant 2. The US graph profits more than the European graph since it has paths with considerably larger hop counts, perhaps due to a larger number of degree two nodes in the input. Note that due to cache effects, the time for outputting the path using preprocessed shortcuts is likely to be considerably smaller than the time for traversing the shortest path in the original graph.

Table 7.9: Additional preprocessing (pp) time, additional disk space and query time that is needed to determine a complete description of the shortest path and to traverse it summing up the weights of all edges—assuming that the query to determine its lengths has already been performed. Moreover, the average number of hops—i.e., the average path length in terms of number of nodes—is given. The three algorithmic variants are described in Section 3.4.3.

	Europe				USA (Tiger)			
	pp [s]	space [MB]	query [ms]	# hops (avg.)	pp [s]	space [MB]	query [ms]	# hops (avg.)
Variant 1	0	0	17.22	1 366	0	0	39.69	4 410
Variant 2	69	126	0.49	1 366	68	127	1.16	4 410
Variant 3	74	225	0.19	1 366	70	190	0.25	4 410

7.4.7 Turning Restrictions

We did an experiment with the German road network (a subgraph of our European network) and real-world turning restrictions (also provided by PTV) to verify our expectation that incorporating the restrictions into the graph has only a little effect on the performance. The results are positive: the preprocessing time does not change, the total number of nodes and edges in the highway hierarchy only increases by 4%, and the query times rise by 3%.

7.4.8 Distance Metric

When we apply a distance metric instead of the usual (and for most practical applications more relevant) travel time metric, the hierarchy that is inherent in the road network is less distinct since the difference between fast and slow roads fades. We no longer observe the self–similarity in the sense that a fixed neighbourhood size yields an almost constant shrinking factor. Instead, we have to use an increasing sequence of neighbourhood sizes to ensure a proper shrinking. For Europe, we use $H = 100, 200, 300, 400, 500$ to construct five levels before an all–pairs distance table is built. Constructing the hierarchy takes 34 minutes and entails a memory overhead of 36 bytes per node. On average, a random query then takes 4.88 ms, settling 4 810 nodes and relaxing 33 481 edges. Further experiments on different metrics can be found in Section 7.5.

7.4.9   An Even Larger Road Network

In order to deal with our new and larger European road network (New Europe), we use the same parameters as for the old version (in particular, $H = 30$). We observe a very good shrinking behaviour: we have 1.87 times as many nodes in the beginning (compared to the old version), but after the construction of the same number of levels only 1.04 times as many nodes remain. Thus, the same number of levels is sufficient, only the distance table gets slightly bigger. We arrive at a preprocessing time of 18 minutes, a memory overhead of 37 bytes per node, and query times of 0.60 ms for random queries; on average, 685 nodes are settled and 2 457 edges are relaxed.

7.5 Highway Hierarchies Combined With Goal-Directed Search

7.5.1 Parameters

Unless otherwise stated, we use the same default settings as specified in Section 7.4.1. The chosen neighbourhood sizes for all considered metrics are given in Table 7.10. We use 16 maxCover landmarks that have been computed in the level–3 core. The approximate query algorithm uses a max–imum relative error of 10%, i.e., $\varepsilon = 0.1$.

Table 7.10: Used neighbourhood sizes. For the travel time metric, we use a fixed neighbourhood size for the construction of all levels. For the other two metrics, we use linearly increasing sequences as neighbourhood sizes of the different levels. For Europe with the travel time metric, we have a different neighbourhood size in case that we do not use a distance table ($\emptyset$) and in case that we use one (DT).

metric	Europe		USA (Tiger)	
	$\emptyset$	DT	$\emptyset$	DT
time	40	30	40	40
dist	60, 120, 180, ...		60, 120, 180, ...	
unit	40, 50, 60, ...		60, 120, 180, ...	

7.5.2 Using a Distance Table and/or Landmarks

As already mentioned in Section 3.4.2, using a distance table can be seen as adding a very strong sense of goal direction after the core of the topmost level has been reached. If the highway query algorithm (without distance table) is enhanced by the ALT algorithm, the goal direction comes into ef–fect much earlier. Still, the most considerable pruning effect occurs in the middle of rather long paths: close to the source and the target, the lower bounds are too weak to prune the search. Thus, both optimisations, distance tables and ALT, have a quite similar effect on the search space: using either of both techniques, in case of the European network with the *travel time metric*, the search space size is roughly halved (see Table 7.11). When we consider other aspects like preprocessing time, memory usage, and query time, we can conclude that the distance table is somewhat superior to the landmarks optimisation. Since both techniques have a similar point of application, a combination of the highway query algorithm with both optimisations gives only a comparatively small improvement compared to using only one optimisation. In contrast to the exact algorithm, the approximate variant reduces the search space size and the query time considerably—e.g., to 19% and 25%, respectively, in case of Europe (relative to using only the distance table optimisation)—, while guaranteeing a maximum relative error of 10% and achieving a total error[10] of 0.051% in our random sample (refer to Table 7.12).

Using a *distance metric*, ALT gets more effective and beats the distance table optimisation since much better lower bounds are produced: the negative effect described in Figure 3.16 is weakened. Furthermore, in this case, a combination with both optimisations is worthwhile: the query time is reduced to 43% in case of Europe (relative to using only the distance table optimisation). While the highway query algorithm enhanced with a distance table has 5.9 times slower query times when applied to the European graph with the distance metric instead of using the travel time metric, the combination with both optimisations reduces this performance gap to a factor of 3.1—or even 1.2 when the approximate variant is used.

The performance for the *unit metric* ranks somewhere in between. Although computing shortest paths in road networks based on the unit metric seems kind of artificial, we observe a hierarchy in this scenario as well, which explains the surprisingly good preprocessing and query times: when we drive on urban streets, we encounter much more junctions than driving on a national road or even a motorway; thus, the number of road segments on a path is somewhat correlated to the road type. It is difficult to tell why the US road network with the unit metric is considerably more difficult to handle than the European network. Originally, we tried using the same neighbourhood sizes for both networks. But it turned out that Europe shrinks much better and that the US network requires larger neighbourhood sizes (cp. Table 7.10), which has a negative impact on the performance.

---

[10]i.e., the sum of the path lengths obtained by the approximate algorithm divided by the sum of the shortest-path lengths minus one

Table 7.11: Comparison of all variants of the highway query algorithm using no optimisation (∅), a distance table (DT), ALT, or both techniques. Values in parentheses refer to *approximate* queries.

metric		∅	DT	ALT	both	
		Europe				
time	preproc. time [min]	13	13	14	14	
	total disk space [MB]	898	1 241	1 301	1 644	
	#settled nodes	1 510	708	786	511	(134)
	query time [ms]	1.19	0.60	0.80	0.49	(0.15)
dist	preproc. time [min]	31	32	33	33	
	total disk space [MB]	907	1 654	1 309	2 056	
	#settled nodes	7 685	3 261	2 445	1 449	(125)
	query time [ms]	6.99	3.53	2.72	1.51	(0.18)
unit	preproc. time [min]	21	22	23	24	
	total disk space [MB]	903	1 358	1 302	1 757	
	#settled nodes	3 015	1 524	1 550	1 116	(645)
	query time [ms]	2.42	1.37	1.55	1.11	(0.68)
		USA (Tiger)				
time	preproc. time [min]	16	16	17	18	
	total disk space [MB]	1 124	1 283	1 649	1 807	
	#settled nodes	1 553	932	803	627	(132)
	query time [ms]	1.04	0.70	0.72	0.55	(0.15)
dist	preproc. time [min]	35	38	37	40	
	total disk space [MB]	1 126	2 139	1 651	2 663	
	#settled nodes	7 461	3 512	2 059	1 372	(117)
	query time [ms]	6.03	3.73	2.16	1.37	(0.20)
unit	preproc. time [min]	36	36	38	38	
	total disk space [MB]	1 108	1 562	1 630	2 083	
	#settled nodes	7 126	3 676	2 781	1 778	(306)
	query time [ms]	5.18	3.15	2.55	1.60	(0.36)

7.5.3 Local Queries

In Figure 7.6, we compare the exact and the approximate HH* search in case of the European network with the travel time metric. In the exact case, we observe a continuous increase of the query times: since the distance between source and target grows, it takes longer till both search scopes meet.

For large Dijkstra ranks, the slope decreases. This can be explained by the distance table that bridges the gap between the forward and backward search for long-distance queries very efficiently (cp. Section 7.4.3).

Up to a Dijkstra rank of 2^{18}, the approximate variant shows a very similar behaviour—even though at a somewhat lower level. Then, the query times *decrease*, reaching very small values for very long paths (Dijkstra ranks 2^{22}–2^{24}). This is due to the fact that the *relative* inaccuracy of the lower bounds, which is crucial for the stop condition of the approximate algorithm, is less distinct for very long paths: hence, most of the time, the lower bounds are sufficiently strong to stop very early. However, the large number and high amplitude of outliers indicates that sometimes goal direction does not work well even for approximate queries.

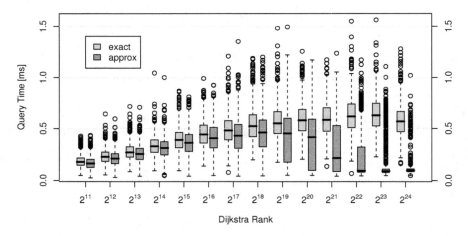

Figure 7.6: Local queries on the European network with the travel time metric using the exact and the approximate HH* search.

7.5.4 Approximation Error

Figure 7.7 shows the actual distribution of the approximation error for a random sample in the European network with the travel time metric. For paths up to a moderate length (Dijkstra rank 2^{16}), at least 99% of all queries in the random sample returned an accurate result. Only very few queries approach the guaranteed maximum relative error of 10%. For longer paths, still more than 93% of the queries give the correct result, and almost 99% of the queries find paths that are at most 2% longer than the shortest path. The fact that we get more errors for longer paths corresponds to the running times depicted in Figure 7.6: in the case of large Dijkstra ranks, we usually stop the search quite early, which increases the likelihood of an inaccuracy.

While the approximate variant of the ALT algorithm [17] gives only a small speedup and produces a considerable amount of inaccurate results (in particular for short paths), the approximate HH* algorithm is much faster than the exact version (in particular for long paths) and produces a compar–atively small amount of inaccurate results. This difference is mainly due to the distance table, which allows a fast determination of upper bounds—and thus, in many cases early aborts—and provides accurate long–distance sub–paths, i.e., the only thing that can go wrong is that the search processes in the local areas around source and target do not find the right core entrance points.

In Table 7.12, we compared the effect of different values for the maxi–mum relative error ε. We obtained the expected result that a larger maximum relative error reduces the search space size considerably. Furthermore, we

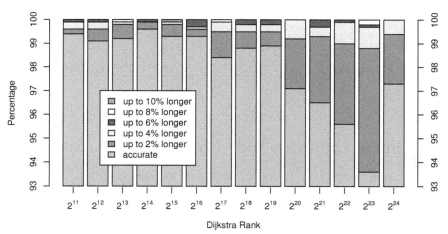

Figure 7.7: Actual distribution of the approximation error for a random sam–ple in the European network with the travel time metric, grouped by Dijkstra rank. Note that, in order to increase readability, the y–axis starts at 93%, i.e., at least 93% of all queries returned an accurate result.

had a look at the actual error that occurs in our random sample: we divided the sum of all path lengths that were obtained by the approximate algorithm by the sum of the shortest path lengths (and subtracted one). We find that the resulting *total error* is *very* small, e.g., only 0.051% in case of the European network with the travel time metric when we allow a maximum relative error of 10%.

Table 7.12: Comparison of different maximum relative errors ε. Note that the observed total errors are given in percent.

metric	ε [%]	0	1	2	5	10	20
		Europe					
time	#settled nodes	511	459	393	241	134	83
	total error [%]	0	0.0002	0.0016	0.015	0.051	0.107
dist	#settled nodes	1 449	851	542	217	125	101
	total error [%]	0	0.0091	0.0351	0.116	0.211	0.278
unit	#settled nodes	1 116	1 074	1 030	879	645	344
	total error [%]	0	0.0001	0.0002	0.004	0.024	0.127
		USA (Tiger)					
time	#settled nodes	627	506	417	249	132	75
	total error [%]	0	0.0025	0.0141	0.062	0.125	0.188
dist	#settled nodes	1 372	754	465	193	117	94
	total error [%]	0	0.0112	0.0302	0.083	0.132	0.166
unit	#settled nodes	1 778	1 450	1 170	636	306	157
	total error [%]	0	0.0010	0.0065	0.044	0.146	0.282

7.6 Static Highway-Node Routing

All results in this section refer to the European road network. We also did some experiments on USA/CAN and USA (Tiger), which indicate that highway-node routing works similarly well on North American networks.

7.6.1 Parameters

We construct a highway hierarchy without a distance table using the parameters specified in Section 7.5.1. We get a classification of the nodes into 12

levels. In order to obtain a variant with an outstanding low memory consumption, we also derive a classification into 11 levels, where level 1 is just
omitted.

After performing a lot of preliminary experiments, we decided to apply
the stall–on–demand technique to the query and the stall–in–advance technique to the construction process (with $a := 5$ except for the construction
of level 1 in the 12–level case, where we use $a := 1$). Moreover, we use the
edge reduction step in order to compute minimal overlay graphs.

7.6.2 Results

Table 7.13 gives an overview on the performance of different variants of
static highway–node routing. We consider the 12–level variant ('normal')
and the one with 11 levels ('save memory'). In each case, we distinguish
between a version where we keep the complete overlay–graph hierarchy and
a version where we keep only the search graphs. Note that the search space
sizes differ between the two versions because the stall–on–demand technique
is slightly less effective when applied only within the search graph since a
few useful edges are missing.

In case of the 11–level variant, level 1 consists of comparatively few
nodes so that, on the one hand, a local search in level 0 takes comparatively
long till the search tree is covered by level–1 nodes. This is reflected in

Table 7.13: Performance of different variants of static highway–node routing. We give both the time to construct the highway hierarchy (HH) that
determines the highway–node sets and the time to construct the multi–level
overlay graph used for highway–node routing (HNR).

	normal		save memory	
	complete	search graph	complete	search graph
constr. HH [min]	11:27			
constr. HNR [min]	3:31		7:44	
overhead [B/node]	9.5	2.4	4.0	0.7
query [ms]	0.89	0.85	1.50	1.44
#settled nodes	957	981	2 328	2 369
#relaxed edges	7 561	7 737	10 693	10 927

slower construction and query times. On the other hand, the first overlay graph is comparatively small so that only little memory is needed for the additional edges. When we keep only the forward and backward search graph, the memory overhead is as little as about 6 *bits* per node (on average).

Even if we consider the normal variant and keep the complete overlay–graph hierarchy, the space overhead is still quite small: less than 10 bytes per node to store the additional edges of the multi-level overlay graph and the level data associated with the nodes. The total disk space[11] of 33.2 bytes per node also includes the original edges and a mapping from original to internal node IDs (that is needed since the nodes are reordered by level).

Details on the 12-level overlay graph can be found in Table 7.14. We observe that the shrinking factor decreases from level to level. This is due to the fact that this particular multi-level overlay graph is based on a highway hierarchy with a rather small neighbourhood size of 40. The average node degree increases from level to level, but stays within reasonable bounds.

Table 7.14: Details on the 12-level overlay graph used for static highway–node routing. Note that the edge counters do *not* include edges that can be only used in a backward search.

level	#nodes	shrink factor	#edges	shrink factor	average degree
0	18 029 721		42 199 587		2.3
1	2 739 732	6.6	11 884 352	3.6	4.3
2	423 635	6.5	2 226 290	5.3	5.3
3	118 844	3.6	780 147	2.9	6.6
4	35 617	3.3	292 630	2.7	8.2
5	11 944	3.0	117 123	2.5	9.8
6	4 364	2.7	49 290	2.4	11.3
7	1 817	2.4	23 108	2.1	12.7
8	864	2.1	12 434	1.9	14.4
9	454	1.9	6 579	1.9	14.5
10	249	1.8	4 029	1.6	16.2
11	146	1.7	2 459	1.6	16.8

---

[11]The main memory usage is somewhat higher. However, we cannot give exact numbers for the static variant since our implementation does not allow to switch off the dynamic data structures.

Figure 7.8 shows the query performance against the Dijkstra rank. The characteristics are similar to those of highway hierarchies (cp. Figure 7.3). For small Dijkstra ranks, highway–node routing is somewhat superior, while for large Dijkstra ranks, highway hierarchies take the lead. The latter observation can be explained by the fact that the current implementation of highway–node routing does not make use of the distance table optimisation (Section 3.4.2).[12]

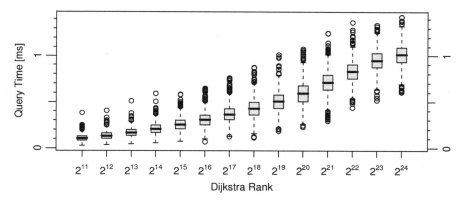

Figure 7.8: Local queries.

Figure 7.9 gives upper bounds for the search space sizes of all possible s–t–queries (analogously to Figure 7.4). We can guarantee that at most 2 148 nodes are settled during any query within the European road network. This is slightly better than the corresponding guarantee that highway hierarchies give (2 388 nodes).

We did a preliminary experiment on the performance of the unidirectional query algorithm (Section 4.4.2). The search space size increased only from 957 to 1 131 nodes. The effects on the query times are not clear yet. On the one hand, a tuned version might take advantage of the fact that we have to manage only a single priority queue. On the other hand, we have to keep in mind that the computation of the reliable levels takes some time as well.

---

[12]An integration of the distance table optimisation would be straightforward. However, it would hinder efficient dynamic updates, which represent the main motivation for the development of highway–node routing. Therefore, we omitted this optimisation.

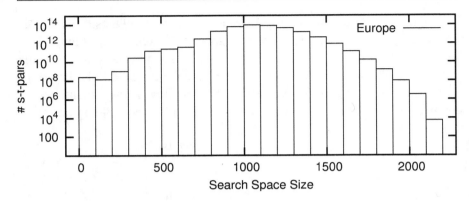

Figure 7.9: Histogram of upper bounds for the search space sizes of s–t–queries.

7.7   Dynamic Highway-Node Routing

As in the previous section, all results in this section refer to the European road network.

7.7.1   Parameters

We deviate from the parameters used in the previous section in order to achieve a different trade–off that favours good update times. In order to determine the highway–node sets, we construct a highway hierarchy consisting of seven levels using the neighbourhood size $H = 70$. This can be done in 16 minutes. For *all* further experiments, these highway–node sets are used.

As before, we use the stall–in–advance technique for the construction and update process (with $a := 1$ for the construction of level 1 and $a := 5$ for all other levels).

7.7.2   Changing the Cost Function

In addition to our default speed profile (introduced in Table 7.2), we consider a few other selected speed profiles (which have been provided by the company PTV AG), namely profiles for a fast car, a slow car, and a slow truck. Table 7.15 gives the construction time of the multi–level overlay graph and the resulting average query performance for all these different

Table 7.15: Construction time of the overlay graphs and query performance for different speed profiles using the same highway-node sets. For the default speed profile, we also give results for the case that the edge reduction step is applied.

speed profile	default	(reduced)	fast car	slow car	slow truck	distance
constr. [min]	1:40	(3:04)	1:41	1:39	1:36	3:56
query [ms]	1.17	(1.12)	1.20	1.28	1.50	35.62
#settled nodes	1 414	(1 382)	1 444	1 507	1 667	7 057
#relaxed edges	9 459	(9 162)	9 746	10 138	11 647	217 857

speed profiles (using the same highway-node sets). Note that for most road categories, our default profile is slightly faster than PTV's fast car profile. The last speed profile ('distance') in Table 7.15 virtually corresponds to a distance metric since for each road type the same constant speed is assumed. The performance in case of the three PTV travel time profiles is quite close to the performance for the default profile. Hence, we can efficiently switch between these profiles without recomputing the highway-node sets.

For the constant speed profile, we get results that are considerably worse (but probably still acceptable in most practical applications). There are two reasons for this observation. First, the constant profile differs significantly from the travel time profiles so that the chosen highway-node sets are no longer as compatible as in the other cases. Second, as already observed in case of highway hierarchies, the constant profile is a more difficult case since the hierarchy inherent in the road network is less distinct. This is confirmed by the fact that when we replace our standard highway-node set with a set that has been determined using the distance metric, both construction and query time are considerably improved to 2:04 minutes and 9.23 ms, respectively, but still the query time cannot compete with the travel time profiles.

We assume that any other 'reasonable' cost function would rank somewhere between our default and the constant profile.

7.7.3 Changing a Few Edge Weights (Server Scenario)

In the dynamic scenario, we need additional space to manage the affected node sets A_u^ℓ. Furthermore, the edge reduction step is not yet supported in the dynamic case so that the total disk space usage increases to 56 bytes per node. In contrast to the static variant, the main memory usage is considerably higher than the disk space usage (around a factor of two) mainly because the dynamic data structures maintain vacancies that might be filled during future update operations.

We can expect different performances when updating very important roads (like motorways) or very unimportant ones (like urban streets, which are usually only relevant to very few connections). Therefore, for each of the four major road categories, we pick 1 000 edges at random. In addition, we randomly pick 1 000 edges irrespective of the road type. For each of these edge sets, we consider four types of updates: first, we add a traffic jam to each edge (by increasing the weight by 30 minutes); second, we cancel all traffic jams (by setting the original weights); third, we block all edges (by increasing the weights by 100 hours, which virtually corresponds to 'infinity' in our scenario); fourth, we multiply the weights by 10 in order to allow comparisons to [18]. For most of these cases, Table 7.16 gives the average update time per changed edge. We distinguish between two change set sizes: dealing with only one change at a time and processing 1 000 changes simultaneously.

As expected, the performance depends mainly on the selected edge and hardly on the type of update. The average execution times for a single update operation range between 40 ms (for motorways) and 2 ms (for urban streets). Usually, an update of a motorway edge requires updates of most levels of the overlay graph, while the effects of an urban-street update are limited to the lowest levels. We get a better performance when several changes are processed at once: for example, 1 000 random motorway segments can be updated in about 8 seconds. Note that such an update operation will be even more efficient when the involved edges belong to the same local area (instead of being randomly spread), which might be a common case in real-world applications.

Table 7.16: Update times per changed edge [ms] for different road types and different update types: add a traffic jam ($+$), cancel a traffic jam ($-$), block a road (∞), and multiply the weight by 10 ($\times$).

|change set|	any road type				motorway			
	$+$	$-$	∞	$\times$	$+$	$-$	∞	$\times$
1	2.7	2.5	2.8	2.6	40.0	40.0	40.1	37.3
1000	2.4	2.3	2.4	2.4	8.4	8.1	8.3	8.1

|change set|	national			regional			urban		
	$+$	$-$	∞	$+$	$-$	∞	$+$	$-$	∞
1	19.9	19.6	20.3	8.4	7.9	8.6	2.1	2.0	2.1
1000	7.1	6.7	7.1	5.3	5.0	5.3	2.0	1.9	2.0

7.7.4 Changing a Few Edge Weights (Mobile Scenario)

In the mobile scenario, we need the same data structures as in the server scenario, in particular the affected node sets A_u^ℓ. In addition, in case of the iterative variant, we need data structures to unpack shortcuts (even if we wanted to determine only the shortest path length). These data structures require additional 43 seconds of preprocessing time and 108 MB of disk space. They have the ability to unpack a shortest path that results from a random query and contains shortcuts in 0.31 ms on average.[13] Moreover, for some technical reasons[14], we need 12 MB to store additional copies of some edges.

Table 7.17 shows for the most difficult case (updating motorways) that using our prudent query algorithm, we can omit the comparatively expen-sive update operation and still get good execution times, at least if only a moderate amount of edge weight changes occur. The iterative variant per-forms clearly better: we see a factor of about 14 in the case of 100 changes.

---

[13]These data structures correspond to 'Variant 2' introduced in Section 3.4.3.

[14]Usually, if an edge belongs to several levels of the overlay graph hierarchy, we store it only once and attach the appropriate level information. The iterative variant, however, requires that the overlay graphs stay completely unchanged when the original graph is mod-ified. Thus, if there is an edge that belongs to the original graph and to some overlay graph, we need two copies so that only the copy in the original graph can be modified.

This is due to the fact that usually very few iterations are sufficient to deter–
mine a shortest path. The single pass variant is robust to different types of
edge weights changes: adding 30 minutes or multiplying the edge weights
by 10 yields virtually the same results. In case of the iterative variant, how–
ever, the number of required iterations depends on the extent of the delays:
for '× 10', we see somewhat better query times (unless there are only very
little changes), which is due to the fact that multiplying the edge weights
by 10 corresponds to an edge weight increase by only about 4 minutes on
average.

Table 7.17: Query performance of the single–pass and the iterative variant
depending on the number of edge weight changes on motorways. For ≤
100 changes, 10 different edge sets are considered; for ≥ 1 000 changes, we
deal only with one set. In the column 'affected queries', we give the average
percentage of queries whose shortest–path length is affected by the changes.

| | | SINGLE PASS | ITERATIVE | | | |
| | affected | query time | query time | #iterations | |
|change set|	queries	[ms]	[ms]	average	max
			+ 30 minutes (× 10)		
1	0.4 %	2.3 (2.3)	1.5 (1.5)	1.0 (1.0)	2 (2)
10	5.7 %	8.5 (8.8)	1.7 (1.7)	1.1 (1.1)	3 (3)
100	40.0 %	47.1 (48.0)	3.4 (3.3)	1.4 (1.4)	5 (5)
1 000	83.7 %	246.2 (244.8)	22.9 (17.6)	2.7 (2.4)	9 (8)
10 000	97.9 %	939.0 (950.9)	492.0 (323.3)	7.9 (6.3)	27 (22)

7.8 Many-to-Many Shortest Paths

All results in this section refer to the European road network.

7.8.1 Implementations and Parameters

We have an implementation based on highway hierarchies and one based on highway-node routing. Both implementations follow the specifications from Section 5.4 quite closely except for the facts that the "accurate backward search" optimisation has not been included yet and the "fewer bucket entries" optimisation is realised only in case of highway-node routing. In order to sort the bucket triples (Section 5.4.3), we just employ the sort routine from the Standard Template Library. We expect that using an implementation that is adapted to the specific situation (e.g., some variant of counting sort) could improve the running times for small distance tables (for large distance tables, the sorting time is insignificant, cp. Figures 7.10 and 7.11). In Table 7.18, we also consider the 'original' implementation by Knopp [51], which is based on highway hierarchies and includes the "fewer bucket entries" optimisation.

For the variant based on highway-node routing, we use a multi-level overlay graph that has been constructed according to the parameters specified in Section 7.6.1. For the experiments using Knopp's implementation, a highway hierarchy has been constructed with a neighbourhood size of 70 and a contraction rate of 1.5. We adopted these settings for our reimplementation that is based on highway hierarchies.

7.8.2 Results

Table 7.18 gives the times needed to compute distance tables of various sizes. Our reimplementation based on highway hierarchies is somewhat slower than Knopp's implementation since we do not employ the "fewer bucket entries" optimisation. We did not made the effort to reimplement this optimisation because it was foreseeable that the variant based on highway-node routing would be superior anyway. As a matter of fact, for large tables, highway-node routing yields almost three times smaller execution times so that we need not much more than a minute to compute a $20\,000 \times 20\,000$ table—in other words, less than $0.2\,\mu s$ per table entry.

Table 7.18: Computing $|S| \times |S|$ distance tables using the implementation based on highway hierarchies (HH) from [51] and our new implementations based on highway hierarchies (HH) and on highway–node routing (HNR). Experiments from [51] have been performed on a similar, but not identical machine. All times are given in seconds.

$|S|$	100	500	1 000	5 000	10 000	20 000
HH [51]	0.2	1.0	2.5	23.8	66.7	211.0
HH	0.6	1.7	3.3	26.3	76.6	247.7
HNR	0.4	0.8	1.4	8.5	23.2	75.1

When comparing ourselves to Dijkstra's algorithm, we have to take into account that, in contrast to Dijkstra's algorithm, we need some preprocessing time—15 minutes in case of highway–node routing. However, the break–even point is already reached for a 77×77 distance table. That means, if we want to compute some table larger than 77×77, it is worth it to invest the preprocessing time even if we want to compute only a single table for the given road network. Note that in many applications, we want to compute more than one table for a given road network so that the advantage of our many–to–many algorithm is even more distinct.

For our implementation based on highway hierarchies and the one based on highway–node routing, Figures 7.10 and 7.11 show how the running times distribute over the four parts of the many–to–many algorithm: back–ward searches, sorting the bucket triples, forward searches, and bucket scanning. For larger distance tables, the time spent for bucket scanning gets dominating so that it is reasonable to choose a smaller topmost level. This effect is more distinct in case of highway hierarchies since highway–node routing causes considerably less bucket entries—as we can see in Table 7.19. In addition to various statistics that confirm our analysis from Section 5.4.1, Table 7.19 provides some experimental results for the many–to–many algorithm based on the original/symmetric variant of highway–node routing. From these data, we can conclude that a direct application of the symmetric variant already yields a reasonable performance, but, obviously, switching to the asymmetric variant brings a significant boost (and has absolutely no disadvantages).

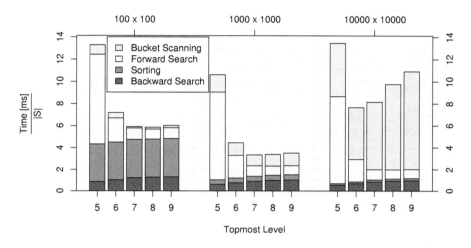

Figure 7.10: Computing $|S| \times |S|$ distance tables using the implementation based on highway hierarchies.

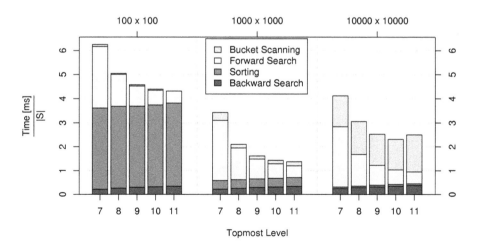

Figure 7.11: Computing $|S| \times |S|$ distance tables using the implementation based on highway-node routing.

Table 7.19: Computing a $10\,000 \times 10\,000$ distance table. For the first three data columns, we give total numbers divided by $10\,000$. We distinguish between the total number of bucket scans ('all') and the number of scans in G'_L or G_L ('top') in case of highway hierarchies or highway–node routing, respectively; in each case divided by $100\,000\,000$. 'Overlap' denotes the quotient of bucket scans and bucket entries: this roughly corresponds to the average overlap of a forward search space and (the non–stalled part of) a backward search space. In case of highway–node routing, we also give results based on the original/symmetric variant.

L	bwd search space size	#bucket entries	fwd search space size	#bucket scans all	(top)	overlap	time [s]
			highway hierarchies				
5	574	574	6 474	64	(61)	11.2%	134.6
6	721	721	2 093	85	(75)	11.8%	76.6
7	884	884	1 154	128	(97)	14.5%	81.6
8	983	983	1 012	172	(81)	17.4%	97.7
9	1 013	1 013	1 008	201	(0)	19.9%	109.2
			highway–node routing				
7	283	168	2 090	17	(15)	10.4%	41.3
8	326	191	1 179	22	(15)	11.6%	30.6
9	356	207	806	23	(19)	11.0%	25.3
10	383	223	637	23	(18)	10.4%	23.2
11	412	238	564	29	(17)	12.3%	25.0
sym	519	359	564	154	(N/A)	42.8%	77.8

7.9 Transit-Node Routing

7.9.1 Parameters

We apply the economical variant to the travel time, the distance, and the unit metric. In each case, in order to determine the highway–node sets (and consequently the transit–node sets) we construct a highway hierarchy us–ing the neighbourhood sizes from Table 7.10 (DT). In addition, we apply the generous variant to the travel time metric using the neighbourhood size $H = 90$.

7.9.2 Main Results

Preprocessing. Table 7.20 gives the preprocessing times for all consid–ered road networks, metrics, and variants. In addition, some key facts on the results of the preprocessing, e.g., the sizes of the transit node sets, are presented. It is interesting to observe that for the travel time metric in level 2 the actual distance table size is at most 0.2% of the size a naive $|T_2| \times |T_2|$ table would have.

Table 7.20: Preprocessing statistics. The size $|D_2|$ of the level–2 distance table is given relative to the size of a complete $|T_2| \times |T_2|$ table. $|A_\ell|$ denotes $|\overrightarrow{A}_\ell \cup \overleftarrow{A}_\ell|$, i.e., the size of the union of forward and backward access nodes.

metric	variant	level 3		level 2			overhead	time
		$|T_3|$	$|A_3|$	$|T_2|$	$|D_2|$	$|A_2|$	[B/node]	[h]
				Europe				
time	eco	9 355	11.4	151 450	0.15%	5.3	99	0:25
	gen	9 458	11.3	293 209	0.14%	4.4	226	1:15
dist	eco	14 001	22.3	179 972	1.03%	8.8	301	2:42
unit	eco	10 923	12.7	212 014	0.28%	6.4	138	0:53
				USA (Tiger)				
time	eco	6 449	6.8	218 153	0.20%	5.2	121	0:38
	gen	10 261	6.1	449 945	0.08%	4.5	257	1:25
dist	eco	16 296	19.1	261 759	0.53%	7.5	280	3:37
unit	eco	10 901	12.5	239 029	1.00%	6.2	219	3:59

As expected, the distance metric yields more access nodes than the travel time metric (a factor 2–3) since not only junctions on very fast roads (which are rare) qualify as access nodes. The fact that we have to increase the neigh- bourhood size from level to level in order to achieve an effective shrinking of the highway networks leads to comparatively high preprocessing times for the distance metric.

Random Queries Using the Travel Time Metric.   Table 7.21 summarises the average case performance of transit–node routing. For the travel time metric, the generous variant achieves average query times more than two orders of magnitude lower than those of highway–node routing (Table 7.13) or highway hierarchies (even when combined with goal–directed search, Ta- ble 7.11). Compared to Dijkstra's algorithm, we obtain a speedup of a factor 1.4 and 1.9 *million* in case of Europe and the USA, respectively. At the cost of about a factor three in query time, the economical variant saves around a factor of two in space and a factor of 2–3 in preprocessing time.

Finding a good locality filter is one of the biggest challenges of our instantiation of transit–node routing. The values in Table 7.21 indicate that our filter is suboptimal: for instance, only 0.0051% of the queries performed by the economical variant in the European network would require a local search to answer them correctly. However, the locality filter $\mathcal{L}_2$ forces us to perform local searches in 0.1364% of all cases. The high–quality level–2 filter employed by the generous variant is considerably more effective, still the percentage of false positives exceeds 90%.

Random Queries Using the Distance or Unit Metric.   For the distance and unit metric, the situation is worse. A considerably larger fraction of the queries continues to level 2 and below. It is important to note that we have concentrated on the travel time metric—since we consider the travel time metric more important for practical applications—, and we spent com- paratively little time to tune our approach for the distance and unit metric. Nevertheless, the current version shows feasibility and still achieves an im- provement of a factor of at least 15 or 80 for the distance or unit metric, respectively, compared to highway hierarchies combined with goal–directed search (Table 7.11).

Table 7.21: Query statistics w.r.t. $10\,000\,000$ randomly chosen (s,t)-pairs. Each query is performed in a top-down fashion. For each level ℓ, we report the percentage of the queries that are not answered correctly in some level $\geq \ell$ and the percentage of the queries that are not stopped after level ℓ (i.e., $\mathcal{L}_\ell(s,t)$ is true). Note that only the generous variant can perform a query in level 1 (but, as the economical variant, it always continues to level 0).

metric	variant	level 3 [%]		level 2 [%] (level 1 [%])		query time
		wrong	cont'd	wrong	cont'd	
			Europe			
time	eco	0.57	3.36	0.0051	0.1364	$11.0\,\mu s$
	gen	0.25	1.55	0.0016 (0.00019)	0.0180 (0.0180)	$4.3\,\mu s$
dist	eco	3.89	13.21	0.0121	0.4897	$37.6\,\mu s$
unit	eco	1.06	5.23	0.0070	0.1731	$13.1\,\mu s$
			USA (Tiger)			
time	eco	0.37	2.44	0.0045	0.1130	$9.5\,\mu s$
	gen	0.10	0.87	0.0010 (0.00009)	0.0124 (0.0124)	$3.3\,\mu s$
dist	eco	1.04	5.35	0.0067	0.1587	$86.1\,\mu s$
unit	eco	1.67	8.66	0.0099	0.2729	$19.8\,\mu s$

Local Queries Using the Travel Time Metric. Since the overwhelming majority of all cases is handled in the top level (more than 99% in case of the US network using the generous variant), the average case performance says little about the performance for more local queries which might be very important in some applications. Therefore, we use the methodology introduced in Section 7.3.2 to get more detailed information about the query time distributions for queries ranging from very local to global, see Figure 7.12. Note that even the median query times for the largest Dijkstra rank (which is the best case) are higher than the average query times given in Table 7.21. This is due to the fact that logging the statistics required to create the depicted plot causes a certain overhead.

For the generous approach, we can easily recognise the three levels of transit-node routing with small transition zones in between: For ranks 2^{18}–

2^{24} we usually have $\neg\mathcal{L}_3(s, t)$ and thus only require cheap distance table accesses in level 3. For ranks 2^{12}–2^{16}, we need additional lookups in the table of level 2 so that the queries get somewhat more expensive. In this range, outliers can be considerably more costly, indicating that occasional local searches are needed. For small ranks we usually need local searches and additional lookups in the level-1 table. Still, the combination of a local search in a very small area and table lookups in all three levels usually results in query times of less than $30\,\mu s$.

In case of the economical approach, we observe a high variance in query times for ranks 2^{13}–2^{14}. In this range, all types of queries occur and the difference between the level-3 queries and the local queries is rather big since the economical variant does not make use of level 1. For small Dijkstra ranks, we see a growth of the query times that is typical for highway-node routing (or highway hierarchies).

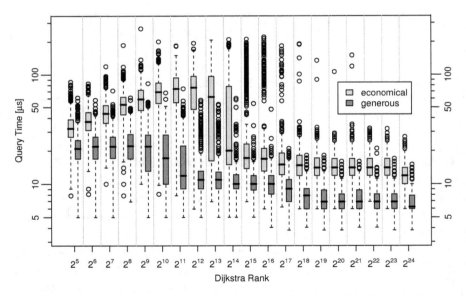

Figure 7.12: Local queries on the European network with the travel time metric using the economical and the generous variant.

Distance Table Accesses. Figure 7.13 represents a histogram of the num‐ber of topmost distance table accesses during an s–t–query. For Europe, we observe an average number of table accesses of 75 and a maximum of $37 \cdot 40 = 1\,480$. Note that these values are by far smaller than the corre‐sponding figures in case of highway hierarchies (Section 7.4.5). This is due to the redefinition of the access nodes mentioned in Section 1.3.5.

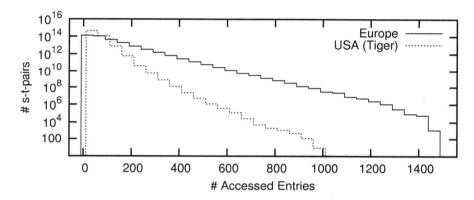

Figure 7.13: Histogram of the number of entries in the topmost distance table that have to be accessed during an s–t–query.

7.9.3 Outputting Complete Path Descriptions

Analogously to Section 7.4.6, in Table 7.22, we report the time that is needed to determine a complete description of the shortest path and to tra‐verse it. We restrict ourselves to the travel time metric and the generous variant. Currently, we provide an efficient implementation only for the case that the path goes through the top level. In all other cases, we just per‐form a normal highway–node query and use the path unpacking routines of highway–node routing. The effect on the average query times is very small since most queries are correctly answered using only the top search. In or‐der to unpack edges of the overlay graphs, we use two different variants that have been introduced as 'Variant 2' and 'Variant 3' in Section 3.4.3. Note that the figures for Variant 3 have been obtained using an older implemen‐tation of transit–node routing based on highway hierarchies and a different set of parameters since the current implementation of highway–node routing does not support this variant.

Table 7.22: Additional preprocessing (pp) time, additional disk space and query time that is needed to determine a complete description of the shortest path and to traverse it summing up the weights of all edges—assuming that the query to determine its lengths has already been performed. Moreover, the average number of hops—i.e., the average path length in terms of number of nodes—is given.

	Europe				USA (Tiger)			
	pp [s]	space [MB]	query [μs]	#hops (avg.)	pp [s]	space [MB]	query [μs]	#hops (avg.)
Variant 2	18	91	314	1 370	29	124	869	4 551
Variant 3	505	227	153	1 370	277	221	264	4 551

7.10 Comparisons

7.10.1 Static Point-to-Point Techniques

In Table 7.23, we compare various variants of our static point-to-point route planning techniques—namely highway hierarchies, static highway-node routing, and transit-node routing—with some of the most competitive methods where experimental results are available for the Western European and the US road network, namely with the REAL algorithm (Section 1.2.4), the edge flag approach (Section 1.2.2), and grid-based transit-node routing (Section 1.2.3). We have to be careful due to several reasons:

- The experiments have been performed on different machines—on different AMD Opteron models, to be more precise. Ours runs at 2.0 GHz, the one used for REAL and the one used for grid-based transit-node routing at 2.4 GHz, and the one used for the edge flag approach at 2.6 GHz. Note that not only the CPU frequency, but also the memory architecture affects the actual machine speed. In fact, our 2.0 GHz machine executes Dijkstra's algorithm slightly faster[15] than the 2.4 GHz machine used for REAL (using the same benchmark implementation).

- Slightly different versions of the road networks are used since some approaches are applied only to the largest strongly connected component

---

[15]about 1.5% on the US road network with travel time or distance metric (cp. Table 7.1 with [31])

(cp. Section 7.2.2). However, this should not lead to any observable dif‐
ferences.

- All approaches allow different choices of parameter settings yielding dif‐
  ferent space‐time trade‐offs. We have to pick a few (particularly reason‐
  able) variants.

- In case of the REAL algorithm, the given memory requirements refer to
  storing landmark distances in a compressed form [32].

- The implementation of the edge flag approach makes use of the Boost
  Graph Library, which seems to cause certain slowdowns. For instance,
  their implementation of bidirectional Dijkstra on the European road net‐
  work takes about 7.2 seconds (on their machine), while ours takes about
  4.4 seconds (on our machine).

- The preprocessing procedure of the edge flag approach explicitly exploits
  the fact that the US network is undirected. If the approach was applied to
  a similar, but directed network, the preprocessing time would double.

- If we are interested not only in the shortest‐path length, but also in a
  complete description of the shortest path, all methods but the edge flags
  have to spend some additional computation time to generate the output.
  For highway hierarchies, highway‐node routing, and transit‐node routing,
  this additional time is given in Table 7.9, Section 7.7.4, and Table 7.22,
  respectively. [31] and [3] state times of about 1 ms and 5 ms for REAL
  and grid‐based transit‐node routing, respectively, to retrieve the shortest
  path (for the US network with the travel time metric).

- The figures for grid‐based transit‐node routing that we quote in Table 7.23
  are based on the most recent, but still *preliminary* version of [3], which
  has been submitted for publication in the final proceedings of the DI‐
  MACS Challenge. We omit the results for the Western European road
  network since they are very tentative.

- The grid‐based implementation of transit‐node routing concentrates on
  global queries. Short‐range queries are comparatively expensive: query
  times of more than 5 ms are reported (while long‐distance queries take
  only 12 μs). Note that accelerating the local queries would require con‐
  siderably more memory.

In spite of these items, we can make some general statements: The strength of our transit–node routing implementation is clearly the extremely good query performance. Highway–node routing has an outstandingly low memory consumption, while the query times are competitive to highway hierarchies and REAL or even slightly superior in case of the US road network. Highway hierarchies can achieve very low preprocessing times or a quite low memory consumption, while query times are reasonably good in all cases. REAL's performance is similar to highway hierarchies except for the preprocessing times, which tend to be considerably higher.

While in Europe the query times of the edge flag approach can keep up with those of other techniques with similar memory requirements, the running times are worse on the US road network. This is due to the facts that the edge flag approach has to visit at least all edges of the shortest path and that an average shortest path in the US network consists of more than three times as many edges as in Europe (cp. Table 7.22). With respect to preprocessing times, edge flags are inferior to most other approaches considered here.

On the one hand, the grid–based transit–node routing has considerably slower preprocessing and query times than our transit–node routing implementation that is based on highway–node routing. On the other hand, the space consumption of the former is much better. However, this is mainly due to the already mentioned fact that the grid–based transit–node routing concentrates only on answering global queries very efficiently.

Table 7.23: Comparison between various *static* route planning techniques: highway hierarchies (HH), highway-node routing (HNR), transit-node routing (TNR), the REAL algorithm, and the edge flag approach (EF). In case of the REAL algorithm, there are two different variants: one with 16 landmarks and one with 64, where landmark distances are kept for $n/16$ highest-reach nodes only. In the first three data columns, the respective best value is highlighted. Rows marked by a ⋆ contain results of additional experiments that have not been published elsewhere.

method	variant	data from	PREPROCESSING		QUERY	
			time [min]	overhead [B/node]	time [ms]	#settled nodes
Europe						
HH	normal	7.4.2	**13**	48.0	0.61	709
HH	save mem	7.4.4	55	17.0	1.10	1 863
HH	+ ALT	7.5.2	14	72.0	0.49	511
HNR	normal	7.6.2	15	2.4	0.85	981
HNR	save mem	7.6.2	19	**0.7**	1.44	2 369
TNR	economical	7.9.2	25	120.0	0.0110	N/A
TNR	generous	7.9.2	75	247.0	**0.0043**	N/A
REAL	16,1	[31]	97	85.0	1.22	814
REAL	64,16	[31]	141	36.0	1.11	679
EF	200 regions	[36]	1 028	19.0	1.6	2 369
EF	1000 regions	[36]	2 156	25.0	1.1	1 593
USA (Tiger)						
HH	normal	7.4.2	**15**	34.0	0.67	925
HH	save mem	⋆	70	17.0	1.21	2 143
HH	+ ALT	7.5.2	18	56.0	0.55	627
HNR	normal	⋆	16	1.6	0.45	784
HNR	save mem	⋆	18	**0.7**	0.61	1 217
TNR	economical	7.9.2	38	143.0	0.0095	N/A
TNR	generous	7.9.2	85	278.0	**0.0033**	N/A
REAL	16,1	[31]	64	109.0	1.14	675
REAL	64,16	[31]	121	45.0	1.05	540
EF	200 regions	[36]	610	10.0	4.3	8 180
EF	1000 regions	[36]	1 419	21.0	3.3	5 522
TNR	grid-based	[3]	900	21.0	0.063	N/A

7.10.2 Dynamic Point-to-Point Techniques

Table 7.24 contains a comparison between dynamic highway-node routing and the dynamic ALT approach [18] (Section 1.2.2) with 16 landmarks. We can conclude that as a stand-alone method, highway-node routing is clearly superior to dynamic ALT w.r.t. all studied aspects.[16]

Table 7.24: Comparison between two *dynamic* route planning techniques: highway-node routing (HNR) and dynamic ALT-16 [18]. Here, 'overhead' denotes the average *disk space* overhead (in bytes per node). Note that highway-node routing—depending on the considered scenario—needs more main memory (see Section 7.7.3). We give execution times for both a complete recomputation using a similar cost function and an update of a single motorway edge multiplying its weight by 10. Furthermore, we give search space sizes after 10 and 1 000 edge weight changes (motorway, ×10) for the mobile scenario. In case of highway-node routing, the iterative variant is used. Time measurements in parentheses have been obtained on a similar, but not identical machine.

method	preprocessing time [min]	overhead	static queries time [ms]	#settled nodes	updates compl. [min]	single [ms]	dynamic queries #settled nodes 10 chgs.	1000 chgs.
HNR	19	39	1.17	1 414	2	37	1 504	17 868
ALT-16	(85)	128	(53.6)	74 441	(6)	(2 036)	75 501	255 754

---

[16]Note that our comparison concentrates on only one variant of dynamic ALT: different landmark sets can yield different trade-offs. Also, better results can be expected when a lot of very small changes are involved. Moreover, dynamic ALT can turn out to be very useful in combination with other dynamic speedup techniques yet to come.

8

Discussion

8.1 Conclusion

While the traditional view on algorithmics is focused almost exclusively on the design and the *theoretical analysis* of algorithms, the paradigm of *algorithm engineering* also includes the implementation and *experimental evaluation* as an essential part of the development process. Using real–world inputs for the conducted experiments is an important ingredient to getting meaningful results. This general statement applies notably to the development of route planning algorithms. Before 2005 only very small road networks were publicly and readily available, which made an evaluation of new techniques under realistic conditions difficult for most researchers.[1] Since then, we have made considerable contributions to obtaining, assembling, and providing (to the scientific community) very large real–world road networks. In particular, our versions of the Western European and the US road network have become the basis of the benchmark instances used at the 9th DIMACS Implementation Challenge [1]. Since it is difficult to obtain a representative list of source–target pairs that originate from real–world applications and since picking source–target pairs just uniformly at random is strongly biased towards very long–distance queries, we introduced a methodology that evaluates a given route planning technique on a

---

[1]The largest real–world road network we had at hand at that time consisted of about 200 000 nodes. However, since each bend was modelled as a distinct node (of degree two), only around 1 000 nodes had a degree greater than two. Thus, the complexity of this network was fairly low.

whole spectrum of queries of different localities. In the meantime, several research groups have adopted both our road networks and our methodology[2] so that now it is comparatively easy[3] to compare results.

In fact, we had the impression that since common standards were available, the race for the best route planning technique has gained momentum, reaching a distinct peak at the DIMACS Implementation Challenge in 2006.[4] In this race, we have taken a leading role. Among all static route planning methods that achieve considerable speedups, we currently provide

- the one with the fastest average query time (transit–node routing, winner of the DIMACS Challenge, 4.3 μs, speedup factor 1.4 million)[5],
- the one with the fastest preprocessing (highway hierarchies, 13 minutes), and
- the one with the lowest memory requirements (highway–node routing, an overhead of 0.7 bytes per node).

In addition, our point–to–point approaches can deal with all types of queries very well, our many–to–many algorithm is unrivalled, and we are not aware of any competitive technique that is able to switch to a different cost function or to handle a moderate amount of traffic jams as efficiently as highway–node routing can do this.

Areas of Application. When dealing with point–to–point queries in a server environment (e.g., route planning systems that provide their services in the internet), transit–node routing can provide excellent response times as long as we consider a static scenario. However, in the case of online route planning systems, transit–node routing might even be an overkill since a significant amount of time is spent on preparing and transmitting graphical representations of the result. Hence, the query times of highway–node

---

[2]sometimes with small modifications

[3]Still, some difficulties remain, in particular the fact that usually different machines are used to run the experiments.

[4]A chronological summary of the 'race' can be found in [72].

[5]Numbers refer to our Western European road network with about 18 million nodes and to our 2.0 GHz machine.

routing would probably be perfectly sufficient. Furthermore, when applying highway–node routing, some dynamic scenarios can be handled as well.

Highway–node routing is also our method of choice when considering a mobile scenario (e.g., a car navigation system). In this case, a concrete real–isation can take advantage of the conceptual simplicity and the low memory requirements.

For some optimisation problems in the field of logistics, a lot of shortest–path queries are required so that transit–node routing can play to its strength. In the special case of a many–to–many problem, our corresponding many–to–many algorithm can be used. A particular example is the (presumably approximate) solution of the travelling salesman problem in a road network, which requires a computation of the shortest paths between the involved nodes.

A direct application of our approaches to traffic simulations is less clear since often time–dependent edge weights have to be considered. The exten–sion to such scenarios is one of the topics that we state in the next section as open questions.

8.2 Future Work

In the concluding remarks of several chapters, we have listed possible fur–ther developments of existing work—some of the mentioned projects have already commenced. In particular, we have started to determine better highway–node sets, to parallelise the preprocessing of highway–node rout–ing, and to write an implementation of highway–node routing for a mobile device. For the first two projects, first promising, though tentative, results are available.

In addition to these concrete advancements, there are various new chal–lenges for next generation route planners that arise from the considerably increasing availability of dynamic road data on the current and the upcom–ing traffic situation and the client's demand for route planning tailored to his individual needs:

Considering Current Traffic Situations.   One challenge is to deal with a *massive* amount of updates to the cost function. These updates reflect the current traffic situation, in particular unexpected events like traffic jams and their effects on the surrounding area. The frequency and extent of these updates will increase significantly over the next few years since not only the coverage of existing monitoring systems like fixed road sensors will be expanded, but also new techniques like floating car data will be widely spread. So far, existing methods like highway–node routing can cope only with a moderate amount of changes.

Considering Upcoming Traffic Situations. Another challenge is to incorporate predictions for upcoming traffic conditions. Such predictions are based on statistical/historical data and are expressed by time–dependent cost functions, which can project, for example, a slower average speed during the morning rush hour. A direct application of existing approaches would fail since a time–expanded representation of a large road network would exceed the available memory. Furthermore, in a time–dependent scenario, all bidirectional search techniques face the problem that simultaneously performing a forward and a backward search normally requires the knowledge of both the exact departure and the exact arrival time.

Multi-Criteria Optimisations. A third challenge is to allow more flexible cost models, dealing with individual compromises between various objective functions like time, financial costs, convenience, environmental pollution, and perhaps scenic value. Interestingly, this topic is related to the problem of dealing with dynamically changing cost functions in the sense that a solution for one problem can turn out to be useful for the other problem as well.

Bibliography

[1] 9th DIMACS Implementation Challenge. Shortest Paths. http://www.dis.uniroma1.it/~challenge9/, 2006.

[2] R. K. Ahuja, K. Mehlhorn, J. B. Orlin, and R. E. Tarjan. Faster algorithms for the shortest path problem. *Journal of the ACM*, 37(2):213–223, 1990.

[3] H. Bast, S. Funke, and D. Matijevic. TRANSIT—ultrafast shortest-path queries with linear-time preprocessing. In *9th DIMACS Implementation Challenge [1]*, 2006.

[4] H. Bast, S. Funke, D. Matijevic, P. Sanders, and D. Schultes. In transit to constant time shortest-path queries in road networks. In *Workshop on Algorithm Engineering and Experiments (ALENEX)*, pages 46–59, 2007.

[5] H. Bast, S. Funke, P. Sanders, and D. Schultes. Fast routing in road networks with transit nodes. *Science*, 316(5824):566, 2007.

[6] R. Bauer. Dynamic speed-up techniques for Dijkstra's algorithm. Diploma Thesis, Universität Karlsruhe (TH), 2006.

[7] R. Bauer and D. Delling. SHARC: Fast and robust unidirectional routing. In *Workshop on Algorithm Engineering and Experiments (ALENEX)*, 2008. To appear.

[8] R. Bauer, D. Delling, and D. Wagner. Experimental Study on Speed-Up Techniques for Timetable Information Systems. In *7th Workshop on Algorithmic Approaches for Transportation Modeling, Optimization, and Systems (ATMOS'07)*. Schloss Dagstuhl, Germany, 2007.

[9] T. Bingmann. Visualisierung sehr großer Graphen. Student Research Project, Universität Karlsruhe (TH), 2006.

[10] U. Brandes, F. Schulz, D. Wagner, and T. Willhalm. Travel planning with self-made maps. In *Workshop on Algorithm Engineering and Experiments (ALENEX)*, volume 2153 of *LNCS*, pages 132–144. Springer, 2001.

[11] U. Brandes, F. Schulz, D. Wagner, and T. Willhalm. Generating node coordinates for shortest-path computations in transportation networks. *ACM Journal of Experimental Algorithmics*, 9(1.1), 2004.

[12] F. Bruera, S. Cicerone, G. D'Angelo, G. Di Stefano, and D. Frigioni. Maintenance of multi-level overlay graphs for timetable queries. In *7th Workshop on Algorithmic Approaches for Transportation Modeling, Optimization, and Systems (ATMOS'07)*. Schloss Dagstuhl, Germany, 2007.

[13] B. V. Cherkassky, A. V. Goldberg, and T. Radzik. Shortest path algorithms: Theory and experimental evaluation. *Math. Programming*, 73:129–174, 1996.

[14] T. H. Cormen, C. E. Leiserson, R. L. Rivest, and C. Stein. *Introduction to Algorithms*. MIT Press, 2nd edition, 2001.

[15] G. B. Dantzig. *Linear Programming and Extensions*. Princeton University Press, 1962.

[16] D. Delling, M. Holzer, K. Müller, F. Schulz, and D. Wagner. High-performance multi-level graphs. In *9th DIMACS Implementation Challenge [1]*, 2006.

[17] D. Delling, P. Sanders, D. Schultes, and D. Wagner. Highway hierarchies star. In *9th DIMACS Implementation Challenge [1]*, 2006.

[18] D. Delling and D. Wagner. Landmark-based routing in dynamic graphs. In *6th Workshop on Experimental Algorithms (WEA)*, 2007.

[19] R. B. Dial. Algorithm 360: Shortest-path forest with topological ordering. *Communications of the ACM*, 12(11):632–633, 1969.

[20] E. W. Dijkstra. A note on two problems in connexion with graphs. *Numerische Mathematik*, 1:269–271, 1959.

[21] J. Fakcharoenphol and S. Rao. Planar graphs, negative weight edges, shortest paths, and near linear time. In *42nd IEEE Symposium on Foundations of Computer Science*, pages 232–241, 2001.

[22] J. Fakcharoenphol and S. Rao. Planar graphs, negative weight edges, shortest paths, and near linear time. *J. Comput. Syst. Sci*, 72(5):868–889, 2006.

[23] I. C. M. Flinsenberg. *Route planning algorithms for car navigation.* PhD thesis, Technische Universiteit Eindhoven, 2004.

[24] M. L. Fredman and R. E. Tarjan. Fibonacci heaps and their uses in improved network optimization algorithms. *Journal of the ACM*, 34(3):596–615, July 1987.

[25] A. Fuchs, M. Mackert, and G. Ziegler. EVA–Netzabbildung und Routensuche für ein fahrzeugautonomes Ortungs– und Navigations–system. *Nachrichtentechnische Zeitschrift*, 36(4):220–223, 1983.

[26] A. Goldberg, H. Kaplan, and R. F. Werneck. Reach for A^*: Efficient point–to–point shortest path algorithms. In *Workshop on Algorithm Engineering and Experiments (ALENEX)*, pages 129–143, Miami, 2006.

[27] A. V. Goldberg. A simple shortest path algorithm with linear average time. In *9th European Symposium on Algorithms (ESA)*, volume 2161 of *LNCS*, pages 230–241. Springer, 2001.

[28] A. V. Goldberg and C. Harrelson. Computing the shortest path: A^* meets graph theory. Technical Report MSR–TR–2004–24, Microsoft Research, 2004.

[29] A. V. Goldberg and C. Harrelson. Computing the shortest path: A^* meets graph theory. In *16th ACM-SIAM Symposium on Discrete Algorithms*, pages 156–165, 2005.

[30] A. V. Goldberg, H. Kaplan, and R. F. Werneck. Better landmarks within reach. In *9th DIMACS Implementation Challenge [1]*, 2006.

[31] A. V. Goldberg, H. Kaplan, and R. F. Werneck. Better landmarks within reach. In *6th Workshop on Experimental Algorithms (WEA)*, volume 4525 of *LNCS*, pages 38–51. Springer, 2007.

[32] A. V. Goldberg and R. F. Werneck. Computing point–to–point shortest paths from external memory. In *Workshop on Algorithm Engineering and Experiments (ALENEX)*, pages 26–40, 2005.

[33] G. Gutin and A. Punnen, editors. *The Traveling Salesman Problem and its Variations.* Kluwer, 2002.

[34] R. Gutman. Reach-based routing: A new approach to shortest path algorithms optimized for road networks. In *Workshop on Algorithm Engineering and Experiments (ALENEX)*, pages 100–111, 2004.

[35] P. E. Hart, N. J. Nilsson, and B. Raphael. A formal basis for the heuristic determination of minimum cost paths. *IEEE Transactions on System Science and Cybernetics*, 4(2):100–107, 1968.

[36] M. Hilger. Accelerating point-to-point shortest path computations in large scale networks. Diploma Thesis, Technische Universität Berlin, 2007.

[37] M. Holzer, F. Schulz, and D. Wagner. Engineering multi-level overlay graphs for shortest-path queries. In *Workshop on Algorithm Engineering and Experiments (ALENEX)*, volume 129 of *Proceedings in Applied Mathematics*, pages 156–170. SIAM, January 2006.

[38] M. Holzer, F. Schulz, and D. Wagner. Engineering multi-level overlay graphs for shortest-path queries. Invited for ACM Journal of Experimental Algorithmics (special issue ALENEX 2006), 2007.

[39] M. Holzer, F. Schulz, and T. Willhalm. Combining speed-up techniques for shortest-path computations. In *3rd International Workshop on Experimental and Efficient Algorithms (WEA)*, volume 3059 of *LNCS*, pages 269–284. Springer, 2004.

[40] T. Ikeda, M.-Y. Hsu, H. Imai, S. Nishimura, H. Shimoura, T. Hashimoto, K. Tenmoku, and K. Mitoh. A fast algorithm for finding better routes by AI search techniques. In *Vehicle Navigation and Information Systems Conference. IEEE*, 1994.

[41] K. Ishikawa, M. Ogawa, S. Azume, and T. Ito. Map Navigation Software of the Electro Multivision of the '91 Toyota Soarer. In *IEEE Int. Conf. Vehicle Navig. Inform. Syst*, pages 463–473, 1991.

[42] R. Jacob and S. Sachdeva. I/O efficieny of highway hierarchies. Technical Report 531, ETH Zürich, Theoretical Computer Science, 2006.

[43] G. Jagadeesh, T. Srikanthan, and K. Quek. Heuristic techniques for accelerating hierarchical routing on road networks. *IEEE Transactions on Intelligent Transportation Systems*, 3(4):301–309, 2002.

[44] N. Jing, Y.-W. Huang, and E. A. Rundensteiner. Hierarchical optimization of optimal path finding for transportation applications. In *5th International Conference on Information and Knowledge Management (CIKM '96)*, pages 261–268. ACM, 1996.

[45] N. Jing, Y.-W. Huang, and E. A. Rundensteiner. Hierarchical encoded path views for path query processing: An optimal model and its performance evaluation. *IEEE Transactions on Knowledge and Data Engineering*, 10(3):409–432, 1998.

[46] S. Jung and S. Pramanik. HiTi graph model of topographical roadmaps in navigation systems. In *12th International Conference on Data Engineering (ICDE '96)*, pages 76–84. IEEE Computer Society, 1996.

[47] S. Jung and S. Pramanik. An efficient path computation model for hierarchically structured topographical road maps. *IEEE Transactions on Knowledge and Data Engineering*, 14(5):1029–1046, 2002.

[48] P. Klein, S. Rao, M. Rauch, and S. Subramanian. Faster shortest-path algorithms for planar graphs. In *26th ACM Symposium on Theory of Computing*, pages 27–37, 1994.

[49] P. N. Klein. Multiple-source shortest paths in planar graphs. In *16th ACM-SIAM Symposium on Discrete Algorithms*, pages 146–155. SIAM, 2005.

[50] S. Knopp. Efficient computation of many-to-many shortest paths. Diploma Thesis, Universität Karlsruhe (TH), 2006.

[51] S. Knopp, P. Sanders, D. Schultes, F. Schulz, and D. Wagner. Computing many-to-many shortest paths using highway hierarchies. In *Workshop on Algorithm Engineering and Experiments (ALENEX)*, 2007.

[52] E. Köhler, R. H. Möhring, and H. Schilling. Acceleration of shortest path and constrained shortest path computation. In *4th International Workshop on Efficient and Experimental Algorithms (WEA)*, 2005.

[53] E. Köhler, R. H. Möhring, and H. Schilling. Fast point–to–point short-est path computations with arc–flags. In *9th DIMACS Implementation Challenge [1]*, 2006.

[54] U. Lauther. An extremely fast, exact algorithm for finding shortest paths in static networks with geographical background. In *Geoinfor-mation und Mobilität – von der Forschung zur praktischen Anwen-dung*, volume 22, pages 219–230. IfGI prints, Institut für Geoinfor-matik, Münster, 2004.

[55] U. Lauther. An experimental evaluation of point–to–point shortest path calculation on roadnetworks with precalculated edge–flags. In *9th DI-MACS Implementation Challenge [1]*, 2006.

[56] R. J. Lipton and R. E. Tarjan. A separator theorem for planar graphs. *SIAM Journal on Applied Mathematics*, 36(2):177–189, April 1979.

[57] J. Maue, P. Sanders, and D. Matijevic. Goal directed shortest path queries using Precomputed Cluster Distances. In *5th Workshop on Experimental Algorithms (WEA)*, number 4007 in LNCS, pages 316–328. Springer, 2006.

[58] K. Mehlhorn and S. Näher. *The LEDA Platform of Combinatorial and Geometric Computing*. Cambridge University Press, 1999.

[59] U. Meyer. Single–source shortest–paths on arbitrary directed graphs in linear average–case time. In *12th Symposium on Discrete Algorithms*, pages 797–806, 2001.

[60] R. Möhring, H. Schilling, B. Schütz, D. Wagner, and T. Willhalm. Partitioning graphs to speed up Dijkstra's algorithm. In *4th Inter-national Workshop on Efficient and Experimental Algorithms (WEA)*, pages 189–202, 2005.

[61] R. Möhring, H. Schilling, B. Schütz, D. Wagner, and T. Willhalm. Partitioning graphs to speed up Dijkstra's algorithm. *ACM Journal of Experimental Algorithmics*, 11(Article 2.8):1–29, 2006.

[62] K. Müller. Berechnung kürzester Pfade unter Beachtung von Abbiege-verboten. Student Research Project, Universität Karlsruhe (TH), 2005.

[63] K. Müller. Design and implementation of an efficient hierarchical speed–up technique for computation of exact shortest paths in graphs. Diploma Thesis, Universität Karlsruhe (TH), 2006.

[64] L. F. Muller and M. Zachariasen. Fast and compact oracles for approx–imate distances in planar graphs. In *15th European Symposium on Algorithms (ESA)*, volume 4698 of *LNCS*, pages 657–668. Springer, 2007.

[65] G. Nannicini, P. Baptiste, G. Barbier, D. Krob, and L. Liberti. Fast paths in large–scale dynamic road networks. arXiv:0704.1068v2 [cs.NI], 2007.

[66] I. Pohl. Bi–directional search. *Machine Intelligence*, 6:124–140, 1971.

[67] R Development Core Team. R: A Language and Environment for Sta–tistical Computing. `http://www.r-project.org`, 2004.

[68] P. Sanders and D. Schultes. Engineering highway hierarchies. Journal version, submitted for publication.

[69] P. Sanders and D. Schultes. Highway hierarchies hasten exact short–est path queries. In *13th European Symposium on Algorithms (ESA)*, volume 3669 of *LNCS*, pages 568–579. Springer, 2005.

[70] P. Sanders and D. Schultes. Engineering highway hierarchies. In *14th European Symposium on Algorithms (ESA)*, volume 4168 of *LNCS*, pages 804–816. Springer, 2006.

[71] P. Sanders and D. Schultes. Robust, almost constant time shortest–path queries in road networks. In *9th DIMACS Implementation Challenge [1]*, 2006.

[72] P. Sanders and D. Schultes. Engineering fast route planning algo–rithms. In *6th Workshop on Experimental Algorithms (WEA)*, volume 4525 of *LNCS*, pages 23–36. Springer, 2007.

[73] D. Schieferdecker. Systematic combination of speed–up techniques for exact shortest–path queries. Diploma Thesis, Universität Karlsruhe (TH), 2007. Under preparation.

[74] W. Schmid. *Berechnung kürzester Wege in Straßennetzen mit Wege-verboten.* PhD thesis, Universität Stuttgart, 2000.

[75] D. Schultes. Fast and exact shortest path queries using highway hier–archies. Master's thesis, Universität des Saarlandes, 2005.

[76] D. Schultes and P. Sanders. Dynamic highway-node routing. In *6th Workshop on Experimental Algorithms (WEA)*, volume 4525 of *LNCS*, pages 66–79. Springer, 2007.

[77] F. Schulz. *Timetable information and shortest paths.* PhD thesis, Uni–versität Karlsruhe (TH), Fakultät für Informatik, 2005.

[78] F. Schulz, D. Wagner, and K. Weihe. Dijkstra's algorithm on-line: An empirical case study from public railroad transport. In *3rd Workshop on Algorithm Engineering (WAE)*, volume 1668 of *LNCS*, pages 110–123. Springer, 1999.

[79] F. Schulz, D. Wagner, and K. Weihe. Dijkstra's algorithm on-line: an empirical case study from public railroad transport. *ACM Journal of Experimental Algorithmics*, 5:12, 2000.

[80] F. Schulz, D. Wagner, and C. D. Zaroliagis. Using multi-level graphs for timetable information. In *Workshop on Algorithm Engineering and Experiments (ALENEX)*, volume 2409 of *LNCS*, pages 43–59. Springer, 2002.

[81] R. Sedgewick and J. S. Vitter. Shortest paths in Euclidean space. *Algorithmica*, 1:31–48, 1986.

[82] J. G. Siek, L.-Q. Lee, and A. Lumsdaine. *The Boost Graph Library: User Guide and Reference Manual.* Addison–Wesley, 2001.

[83] S. S. Skiena. *The Algorithm Design Manual.* Springer, 1998.

[84] M. Thorup. On RAM priority queues. In *7th ACM-SIAM Symposium on Discrete Algorithms*, pages 59–67, 1996.

[85] M. Thorup. Undirected single source shortest paths in linear time. In *Foundations of Computer Science*, 1997.

[86] M. Thorup. Undirected single source shortest paths in linear time. *Journal of the ACM*, 46(3):362–394, 1999.

[87] M. Thorup. On RAM priority queues. *SIAM Journal on Computing*, 30:86–109, 2000.

[88] M. Thorup. Compact oracles for reachability and approximate distances in planar digraphs. In *42nd IEEE Symposium on Foundations of Computer Science*, pages 242–251, 2001.

[89] M. Thorup. Integer priority queues with decrease key in constant time and the single source shortest paths problem. In *35th ACM Symposium on Theory of Computing*, pages 149–158, 2003.

[90] M. Thorup. Compact oracles for reachability and approximate distances in planar digraphs. *Journal of the ACM*, 51(6):993–1024, 2004.

[91] M. Thorup. Integer priority queues with decrease key in constant time and the single source shortest paths problem. *Journal of Computer and System Sciences*, 69(3):330–353, 2004.

[92] U.S. Census Bureau, Washington, DC. UA Census 2000 TIGER/Line Files. http://www.census.gov/geo/www/tiger/tigerua/ua_tgr2k.html, 2002.

[93] P. van Emde Boas, R. Kaas, and E. Zijlstra. Design and implementation of an efficient priority queue. *Math. Syst. Theory*, 10:99–127, 1977.

[94] D. Wagner and T. Willhalm. Geometric speed-up techniques for finding shortest paths in large sparse graphs. In *11th European Symposium on Algorithms (ESA)*, volume 2832 of *LNCS*, pages 776–787. Springer, 2003.

[95] D. Wagner and T. Willhalm. Drawing graphs to speed up shortest-path computations. In *Workshop on Algorithm Engineering and Experiments (ALENEX)*, 2005.

[96] D. Wagner, T. Willhalm, and C. Zaroliagis. Dynamic shortest path containers. In *3rd Workshop on Algorithmic Methods and Models for*

Optimization of Railways (ATMOS'03), volume 92 of *Electronic Notes in Theoretical Computer Science*, pages 65–84. Elsevier, 2004.

[97] I-L. Wang. *Shortest Paths and Multicommodity Network Flows.* PhD thesis, Georgia Inst. Tech., 2003. http://ilin.iim.ncku.edu.tw/ilin/phdthesis/ilinth2_all.pdf.

[98] T. Willhalm. *Engineering Shortest Path and Layout Algorithms for Large Graphs.* PhD thesis, Universität Karlsruhe (TH), Fakultät für Informatik, 2005.

[99] J. W. J. Williams. Heapsort. *Communications of the ACM*, 7:347–348, June 1964.

A

Implementation

An exhaustive description of every single aspect of the implementation of our route planning techniques would go beyond the scope of this thesis. Thus, after some rather general statements in Section 7.1, here, we want to focus on some particularly interesting data structures and give some details on the respective realisation.

A.1 Graph Data Structures

Although highway hierarchies and highway–node routing are closely related so that it would be possible to design a common graph data structure, we distinguish between two separate implementations, mainly due to the fact that we had to introduce a more flexible graph representation for highway–node routing in order to allow updates of the multi–level overlay graph. Both implementations share a common interface so that many of our algorithms can work on both graph types.

A.1.1 Highway Hierarchies

The graph is represented as *adjacency array*, which is a very space–efficient data structure that allows fast traversal of the graph. There are two arrays, one for the nodes and one for the edges. The edges (u, v) are grouped by the source node u and store only the ID of the target node v and the weight $w(u, v)$. Each node u stores the index of its first outgoing edge in the edge array. In order to allow a search in the backward graph, we have to store an edge (u, v) also as backward edge (v, u) in the edge group of node v. In order to distinguish between forward and backward edges, each edge has a forward and a backward flag. By this means, we can also store two-way edges $\{u, v\}$ (which make up the large majority of all edges in a real-world road network) in a space-efficient way: we keep only one copy of (u, v) and one copy of (v, u), in each case setting both direction flags.

The basic adjacency array has to be extended in order to incorporate the level data that is specific to highway hierarchies. In addition to the index of the first outgoing edge, each node u stores its level-0 neighbourhood radius $r_0(u)$. Moreover, for each node u, all outgoing edges (u, v) are grouped by their level $\ell(u, v)$. Between the node and the edge array, we insert another layer: for each node u and each level $\ell > 0$ that u belongs to, there is a *level node* u_ℓ that stores the radius $r_\ell(u)$ and the index of the first outgoing edge (u, v) in level ℓ. All level nodes are stored in a single array. Each node u keeps the index of the level node u_1. Figure A.1 illustrates the graph representation.

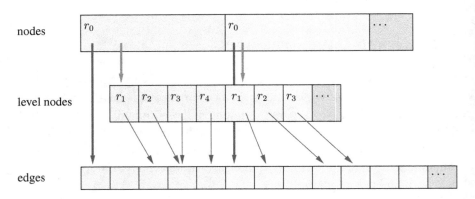

Figure A.1: An adjacency array, extended by a level-node layer.

During construction of the highway hierarchies, we use a variant of this graph data structure where the level nodes are managed in linked lists (since the number of level nodes per node is not known in advance). After the construction has been completed, we can put all level nodes in the right order in the array that then represents the level-node layer.

A.1.2 Highway-Node Routing

The graph data structure used for highway-node routing is very similar to the one used for highway hierarchies. Here, we list the most important differences:

- For each node, there is a level node for level 0—in contrast to highway hierarchies, where the level-0 data is incorporated in the main node.

- The nodes are grouped by level. In contrast to highway hierarchies, this is easily possible because the node levels are known in advance (since the highway-node sets are considered to be part of the input). This allows dealing with nodes that belong to a certain level without the need of scanning all nodes. Furthermore, we do not need to store for each node the index of the first level node. Instead, it is sufficient to store only the respective index i for the very first node u of each level ℓ. Then, for any other node v in level ℓ, the index j of the first level node can be easily computed: $j = i + (v - u) \cdot (\ell + 1)$, exploiting the fact that each node in level ℓ has the same number of level nodes (namely $\ell + 1$). Note that in this equation, we use u and v to denote node IDs in the range $0..n - 1$.

- Obviously, in case of highway-node routing, we do not need to store neighbourhood radii. Thus, the level node of u for level ℓ contains only the last index of the level-ℓ edge group of u in the edge array. We can save some memory by storing only an offset that is added to the first edge index of u.

- In a multi-level overlay graph, an edge e belongs to some consecutive range $k..\ell$ of levels, i.e., $e \in E_k \cap E_{k+1} \cap \ldots \cap E_\ell$. This property has been formally proven in [12]. It is reasonable to store an edge that belongs to several levels $k..\ell$ only once. We put it into the level-ℓ edge group. For performing queries, only this maximum level ℓ (which we just call *level* $\ell(e)$ of e) is relevant. For performing updates, however, we are also interested in the minimum level k (which we also call *creation level* since this is the level where the edge has been created; after that, it has only been upgraded to higher levels). Therefore, we explicitly store k at each edge.

- Most importantly, we allow the addition and deletion of edges at any time. Deletion is comparatively simple: we fill the emerging hole by the last edge in the same level, which leaves a new hole, which is, in turn, filled by the last edge of the next level, and so on; of course, the level-node data has to be updated accordingly. In order to allow efficient additions as well, we ensure that if a node u has x edges, its edge group has a capacity of at least[1] $\min\{2^y \mid 2^y \geq x\}$, i.e., we reserve some space for additional edges. Note that for level-0 nodes, we do not need to reserve additional space since their edge groups never change. Now, adding an edge is straightforward provided that the capacity is not exceeded—we just have to move edges of higher levels to make room at the right spot for the new edge. If, however, the capacity is exceeded, we copy the whole edge group to the end of the edge array (which is, in fact, a resizable STL vector) and double its capacity. Of course, the first edge index of u has to be updated accordingly. Note that these memory management strategies employed by our flexible graph data structure are similar to those used by an STL vector.

A.2 Miscellaneous Data Structures

A.2.1 Priority Queue

Specification. Manages a set of elements with associated totally ordered priorities and supports the following operations:

- *insert* – insert an element,
- *deleteMin* – retrieve the element with the smallest priority and remove it,
- *decreaseKey* – set the priority of an element that already belongs to the set to a new value that is less than the old value.

See also Sections 1.2.1 and 2.2.

Used by all variants of Dijkstra's algorithm.

Implementation. We cannot use the priority queue implementation that the Standard Template Library provides since the *decreaseKey* operation is not supported. Therefore, we use our own straightforward *binary heap*

---

[1]The capacity can be even higher if edge deletions have taken place. This is due to the fact that the capacity is never reduced.

implementation. We have already mentioned in Section 1.2.1 that using a more sophisticated priority queue implementation is unlikely to bring any significant speedup. We did a few preliminary experiments with a 4-ary heap and found an improvement of only 3% for Dijkstra's algorithm. It can be expected that the improvement would be even much lower when applied to one of our route planning techniques where the priority queue operations are less dominant.

A.2.2 Multiple Vector

Specification.   A (large) fixed-sized array of resizable arrays.

Used by the contraction algorithm of highway hierarchies to temporarily store shortcuts and by highway-node routing to store the affected node sets.

Implementation.   Of course, we could just use an STL vector of vectors. However, if we massively add and remove elements from these vectors, we incur serious memory fragmentation and waste. It is important to note that in our applications, such a 'multiple vector' consists of several million vectors. Alternatively, we could use an array of linked lists, which, however, would not be very efficient, either.

We prefer an implementation that is somehow a combination of the two just mentioned possibilities. We manage a vector of data blocks; each block can contain a small fixed number of elements. The blocks can be linked. In the beginning, all allocated blocks are 'free'. We keep a free list of all free blocks. Now, instead of using an array of vectors or an array of linked lists, we employ an array of linked blocks. When an element is added and the current block is full, a new block is requested from the free list and appended. Similarly, empty blocks can be returned. The advantage over a plain linked list implementation is that we do not have to follow a pointer for each single element, which can cause a lot of cache misses. The advantage over a vector of vectors implementation is that the memory overhead is restricted to the number of vectors times the (small) block size.

A.2.3 Multiple Hash Map

Specification. A (large) fixed-sized array of static hash maps.

Used by transit-node routing to store all distance tables but the topmost one.

Implementation. For transit-node routing, we keep large *partial* distance tables, i.e., for a transit-node set $T_\ell, \ell < L$, we have to store a *subset* of the distances $\{d(s,t) \mid (s,t) \in T_\ell^2\}$ (cp. Section 6.2). The data structure should allow fast access times and should be space-efficient. We decided to use for each node $s \in T_\ell$ a hash map that maps a potential target $t \in T_\ell$ to the distance $d(s,t)$. Of course, we could use the hash maps that are part of the STL TR1 extension. However, we are aiming at a more space-efficient solution, which is possible since we can exploit several application-specific properties: the hash maps are static, i.e., we can build them once and for all; many distances are very small since we want to store only the distances that cannot be obtained using higher levels of transit-node routing; the distribution of the node IDs allows the usage of a very simple hash function (namely the least significant bits); two close nodes typically have similar IDs, i.e., the difference of the node IDs is small.

Conceptually, we manage for each node $s \in T_\ell$ its own hash map with chaining that maps t to the distance d from s to t. However, the actual representation is a bit unusual: all hash maps are kept in three common arrays without using any linked lists, as illustrated in Figure A.2.

Let us consider an arbitrary but fixed node $s \in T_\ell$ and assume that the map for s should contain y entries—since we are dealing with static hash maps, this number y is known before we construct the hash map. We compute $x := \lfloor \log_2(y) \rfloor$ and store it in the *main* array at index s. We keep 2^x buckets for s and use the x least significant bits of t as hash function: we denote the hash of t by h. The elements of a bucket are not stored in a linked list, but they are placed one after the other in the so-called *data* array. Since there is only a single data array that contains all elements of all buckets, we need an index structure that allows accessing the first entry of a particular bucket. For this purpose, we have an *index* array. A consecutive range of this array, consisting of 2^x entries, represents the buckets for s. The beginning a of this range is stored in the main array (in addition to x). The sum of a and h is used to address the index array, which contains the index c of the first entry of the corresponding bucket in the data array *relative* to

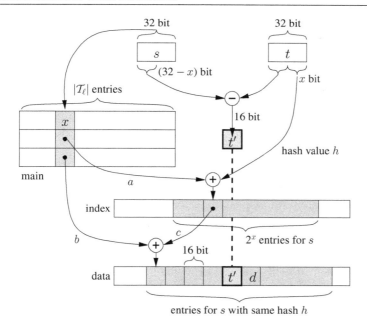

Figure A.2: Using a multiple hash map to look up the distance d from s to t.

the index b of the first entry of the very first bucket of s. The index b is stored in the main array as well so that we can easily compute the required index $b + c$.

Since a bucket can contain several elements with the same hash h but different keys t, we have to store not only the value d, but also the key t so that we can scan the bucket, compare the keys and return the right value. At this point, we want to exploit some of the facts mentioned above to get a very space-efficient implementation. Originally, s, t, and d are 32-bit values. Note that we need not store the x least significant bits of t since all keys in the bucket agree on these bits anyway. Furthermore, in many cases the $(32 - x)$ most significant bits of s and t are very similar so that the difference t' gets very small. It is sufficient to store only this difference, which in most cases requires no more than 16 bits. Often, the value d is so small that it fits in 16 bits as well. Therefore, our data array consists of 16-bit entries, and we store the compressed key t' and the corresponding value d in two consecutive entries. In the exceptional case that t' requires more than 16 bits, we fall back on an additional out-of-the-box hash map (which is not explicitly depicted in Figure A.2) that maps the pair (s, t) to d. In the case of the European road network and the generous variant of transit-node routing, this exceptional case applies to only 0.00011% of all level-2 table entries.

In the other exceptional case that t' is sufficiently small, but d is too large, we store an escape value instead of d and use the next two data entries to represent d.

For the same example as above (Europe, generous, level-2 table), using standard hash maps with chaining would require at least[2] 1 352 MB, while using our multiple hash map occupies only 645 MB, which is less than 50%.

A.2.4   Fast Set

Specification.   Manages a set of integers from a not too large range. Should be optimised for speed, not for space efficiency.

Used by the update procedure of highway-node routing to temporarily represent the set of nodes where the preprocessing step should be repeated from.

Implementation.   Let us assume that the set contains only integers from $0..(k - 1)$. We represent the set by a bit vector of size k, which has the property that the i-th bit is set iff i belongs to the set, and an additional element vector that explicitly stores the elements. Checking whether an element belongs to the set can be done in constant time using the bit vector. If an element i should be inserted, we check whether it already belongs to the set. If not, we set the i-th bit and add i to the element vector, which can be done in amortised constant time. Scanning through all elements can be done using the element vector in time linear in the size of the set.

---

[2]We disregard empty buckets.

A.2.5 Fast Edge Expander

Specification.   Provides data structures to unpack shortcut edges, i.e., to determine the paths in the original graph that correspond to the shortcuts.

Used by highway hierarchies, highway-node routing, and transit-node routing.

Implementation.   As already mentioned in Section 3.4.3, we do not store a sequence of node (or edge) IDs to describe a path, but we store hop indices, i.e., for each edge $e = (u, v)$ on the path, we store $(e - f)$, where f denotes the ID of u's first edge. We put all hops of all represented paths into one large *hops* vector, which consists of 4-bit entries. Since the degree of most nodes is quite small, one such entry is usually sufficient to hold a hop index. In exceptional cases, we write an escape value and use more than one entry to store the hop index.

We need an index structure to access the first entry of the hop sequence that we want to read. Note that we do not need an additional pointer to the end of the sequence since we know that we have gotten to the end as soon as the target of the shortcut edge has been reached. Since not all edges are shortcut edges, it would be wasteful to build an index with m entries. Instead, we use a multi-level index, as depicted in Figure A.3.

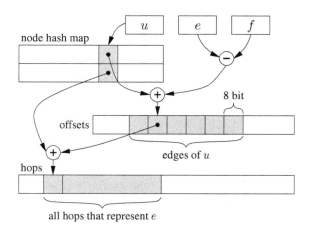

Figure A.3: Using the fast edge expander to access the hops that represent the shortcut edge $e = (u, v)$. The ID of u's first edge is denoted by f.

We keep a node hash map that contains entries only for nodes that have outgoing shortcuts. Only for edges of such nodes, we store an index for the hops vector. However, since these indices are very similar for all edges of the same node u, we store only small *offsets* and add the index of the first hop of u's first shortcut edge to obtain the actual index. The offsets vector consists of 8-bit entries. If an offset does not fit, we store an escape value and use an additional hash map to store the direct mapping from the ID of the shortcut to the right index in the hops vector. Note that this exception hash map is not included in Figure A.3.

Optionally, we use additional, quite similar data structures to store the complete, non-recursive descriptions of the shortcuts that belong to the top-most level (cp. Section 3.4.3, Variant 3).

A.2.6 Search Spaces

Specification. A space-efficient representation of search spaces.

Used by transit-node routing during preprocessing. Could also be used by the many-to-many algorithm.[3]

Implementation. As mentioned in Section 5.4.3, during the backward searches, we manage a single resizable array representing the set $\{(u, t, d) \mid t \in T \wedge (u, d) \in \overleftarrow{\sigma}(t)\}$. In addition to a vector that stores its elements (u, t, d) in three 32-bit integers, we have one vector whose elements consist of one 32-bit and two 16-bit integers. Similarly to Section A.2.3, we want to exploit the fact that the difference of the IDs of u and t is often quite small. Thus, if possible, we store $(u, t - u, d)$ in the more compact vector; if not, we use the normal vector. The order of the search space elements is irrelevant. Therefore, when processing the search spaces later, we just scan the two vectors one after the other.

---

[3]In the current version of the many-to-many implementation, the compression features are switched off since the size of the search spaces is negligible when compared to the size of the distance table.

A.2.7   Access Nodes

Specification.   A space-efficient representation of access nodes.

Used by transit-node routing.

Implementation.   Basically, we use a data structure that is very similar to an adjacency array, i.e., we have one large data vector containing all the access node information and one index vector that contains for each node u the index of the first access node in the data vector. For two nodes $u < v$, it is not always true that the access nodes of u precede the access nodes of v in the data vector.[4] Therefore, in such cases, we cannot use the index stored for the successor node to determine the end of the access node sequence. Instead, we use an explicit end marker, i.e., a value that is not used by any regular access node.

The data vector consists of 16-bit entries. We have to keep the ID of the access node and the distances to and from the access node. We do not store the ID w.r.t. the original graph, but we have a list of all transit nodes and we store only an index within this list. A level-3 access node is stored using 14 bits, a level-2 access node using $16 + 14$ bits, which limits the maximum number of transit nodes in the current implementation.[5] In both cases, we have two remaining bits to indicate whether this particular access node can be used only in forward or backward direction or in both directions, and in the latter case, whether both directions share the same distance. Depending on these direction flags, we store one or two distances; if possible, using one 16-bit entry each; if not, writing an escape value and using two 16-bit entries each.

When we apply the generous variant of transit-node routing to the European road network, storing the level-3 access nodes naively[6] would take 2 512 MB. With our space-efficient representation, however, we need only 1 101 MB.

---

[4]This is due to the fact that usually, we do not determine the access nodes of node 0, then the access nodes of node 1, and so on, but we determine the access nodes of the nodes that belong to a certain transit-node set (which may have arbitrary node IDs) and then hand the access nodes down to all other nodes.

[5]Of course, this restriction could be easily changed.

[6]i.e., forward and backward access nodes separately, 32 bit per access node ID and 32 bit per distance

List of Notation

Zusammenfassung

Die Bestimmung einer optimalen Route in einem Straßennetz von einem gegebenen Start–zu einem gegebenen Zielpunkt ist ein Problem, das viele Menschen täglich beschäftigt. Als Hilfsmittel werden mittlerweile verbreitet Navigationsgeräte eingesetzt oder die Routenberechnung findet im Voraus am Computer statt, beispielsweise unter Verwendung eines der zahlreichen im Internet verfügbaren Dienste. Neben der Routenplanung für den einzelnen PKW gibt es weitere wichtige Anwendungen beispielsweise im Bereich der Logistik.

Es ist naheliegend, ein Straßennetz als Graphen zu repräsentieren. Dabei entspricht eine Straßenkreuzung einem Knoten und eine Straßenabschnitt einer Kante. Aus Sicht der Graphentheorie handelt es sich dann bei der Routenplanung um das sogenannte *kürzeste Wege Problem*. Wir betrachten zwei Varianten: die Berechnung des kürzesten Weges von einem Start–zu einem Zielpunkt und – für gegebene Knotenmengen S und T – die Berechnung einer Distanztabelle, die für jedes Knotenpaar $(s, t) \in S \times T$ die Länge des kürzesten Weges enthält. Prinzipiell könnten wir für beide Problemvarianten auf die 'klassische' Lösung aus der Graphentheorie zurückgreifen, den Algorithmus von Dijkstra. Für große Straßennetze wie beispielsweise das von Westeuropa mit ca. 18 Millionen Straßenkreuzungen wäre dieses Verfahren allerdings für viele praktische Anwendungen zu langsam. Kommerzielle Anbieter setzen daher vielfach schnelle, heuristische Verfahren ein, die darauf verzichten, optimale Routen zu berechnen. Dies hat nicht nur offensichtliche Nachteile für den Benutzer, sondern auch für die Entwickler, da bei jeder Änderung des Programms aufwendig geprüft werden muss, ob sich die Qualität der berechneten Routen noch in einem gewissen Rahmen bewegt.

Aus diesen Gründen besteht ein großes Interesse an exakten und schnellen Routenplanungstechniken. Ein Grundansatz ist hierbei, zunächst etwas Zeit in einen einmaligen Vorberechnungsschritt zu investieren, um Hilfsdaten zu erzeugen, die dann bei allen Routenplanungsanfragen verwendet werden können, um schnelle Suchzeiten zu erreichen. Um auch mit großen Straßennetzen unter Einsatz von begrenzten Resourcen umgehen zu können, sollten sowohl der Vorberechnungsaufwand als auch der benötigte Speicherplatz für die Hilfsdaten möglichst klein sein. Darüber hinaus wird

angestrebt, Verfahren zu entwicklen, die mit dem gesamten Spektrum an
möglichen Anfragen gut zurecht kommen, also sowohl mit lokalen Anfragen innerhalb der gleichen Stadt als auch mit Routenberechnungen quer
durch einen Kontinent. Des Weiteren ist eine gewisse Flexibilität wün-
schenswert: Dazu gehören das Einbeziehen von unerwarteten Ereignissen
wie beispielsweise Staus oder der Wechsel des Geschwindigkeitsprofils, um
optimale Routen für verschiedene Fahrzeugtypen berechnen zu können.

In dieser Arbeit stellen wir drei verschiedene beweisbar korrekte und
effiziente Verfahren für die Punkt-zu-Punkt Berechnug vor – alle mit un-
terschiedlichen Vorzügen – und ein generisches Verfahren zur Distanzta-
bellenberechnung. Dabei folgen wir dem Ansatz des *Algorithm Engineering*: Neben den traditionellen Aspekten der Algorithmenentwicklung,
dem Entwurf und der theoretischen Analyse, umfasst dieser Ansatz auch
die Implementierung und die experimentelle Auswertung als wesentliche
Bestandteile des Entwicklungsprozesses, den man als Kreislauf auffassen
kann, bei dem experimentelle Ergebnisse neue Impulse für die Verbesserung
des entworfenen Algorithmus liefern können. Die Auswertung erfolgt in
Form einer umfangreichen experimentellen Studie, bei der reale Straßennetze mit vielen Millionen Straßenkreuzungen zum Einsatz kommen. Dabei
betrachten wir nicht nur durchschnittliche Suchzeiten, sondern beschäfti-
gen uns auch mit Anfragen mit unterschiedlichem Schwierigkeitsgrad, be-
stimmen obere Suchraumschranken für gegebene Straßennetze, und führen
Vergleiche zwischen verschiedenen Routenplanungstechniken durch. Im
Einzelnen haben wir die folgenden Verfahren entwickelt.

Highway Hierarchien. Während der Algorithmus von Dijkstra keinerlei
spezielle Annahmen über den Graphen macht, nutzen wir gezielt Eigen-
schaften realer Straßennetze aus. Eine solche Eigenschaft ist eine vorhan-
dene Hierarchie der Straßen: Manche Straßen werden nur von lokalen An-
wohnern benötigt, um ihr Wohngebiet zu verlassen, manche Straßen sind
wichtige Verbindungen zwischen verschiedenen Stadtteilen und manche
Straßen werden sogar für Fernverbindungen benötigt. In einem Vorverar-
beitungsschritt berechnen wir eine feinkörnige Klassifizierung aller Straßen,
die der Routenplanungsalgorithmus dann ausnutzen kann. Es handelt sich
dabei um eine Anpassung der bidirektionalen Variante des Algorithmus von
Dijkstra, die den Suchraum deutlich einschränkt: Mit zunehmender Entfer-

nung von Start und Ziel müssen nur noch wichtigere Straßen betrachtet werden, um immer noch beweisbar optimale Ergebnisse zu erhalten. Eine Kombination der Highway Hierarchien mit zielgerichteter Suche führt zu einer Reduktion der Suchzeiten, die insbesondere dann nennenswert ist, wenn man sich ausnahmsweise mit Näherungslösungen begnügt oder wenn man der Suche eine Distanzmetrik zugrunde legt anstatt der üblichen Reisezeitmetrik.

Highway-Node Routing ist ein mit den Highway Hierarchien verwandtes, bidirektionales und hierarchisches Verfahren. Es ist konzeptionell sehr einfach und unterstützt die schnelle Aktualisierung der vorberechneten Daten, um auf Kantengewichtsänderungen zu reagieren.

Transit-Node Routing basiert auf folgender Beobachtung: Wenn man einen weit entfernten Zielpunkt ansteuert, wird man seinen Startpunkt immer über einen von wenigen wichtigen Verkehrsknotenpunkten verlassen. Am Beispiel von Karlsruhe könnten dies die Auffahrten auf die A 5 und die Rheinbrücke sein. Wenn man zum einen die Reisezeiten von allen Punkten zu den zugehörigen wichtigen Verkehrsknotenpunkten und zum anderen die Reisezeiten zwischen allen wichtigen Verkehrsknotenpunkten berechnet und speichert, kann man eine Reisezeitanfrage zwischen zwei hinreichend entfernten Knoten auf wenige Tabellenzugriffe reduzieren. Um auch Anfragen zwischen lokalen Knotenpaaren effizient beantworten zu können, werden weitere Schichten des gleichen Ansatzes benötigt. Die Bestimmung der wichtigen Verkehrsknotenpunkte der verschiedenen Schichten übernimmt hierbei der Konstruktionsalgorithmus der Highway Hierarchien.

Distanztabellen. Bei unserem Verfahren zur Distanztabellenberechnung handelt es sich um einen generischen Algorithmus, der auf verschiedene Weisen instantiiert werden kann, beispielsweise basierend auf den Highway Hierarchien oder auf Highway–Node Routing. Unsere Methode ermöglicht die Berechnung einer vollständigen $|S| \times |T|$ Distanztabelle und führt dazu im Wesentlichen lediglich $|S|$ Vorwärts– plus $|T|$ Rückwärtssuchen aus anstelle von $|S|$ mal $|T|$ bidirektionalen Suchen.

Bewertung. Das Thema "Routenplanung in Straßennetzen" ist in den letz-
ten Jahren in der Forschung heiß umkämpft: Zahlreiche Verfahren kon-
kurrieren miteinander. Nach unserem Kenntnisstand waren wir die ersten,
die die Straßennetze Westeuropas und der USA, bestehend aus ca. 18 bzw.
24 Millionen Knotenpunkten, vollständig und effizient verarbeiten konnten.
Bei einer Beurteilung der Leistungsfähigkeit betrachtet man in der Regel die
Suchzeiten, die Vorberechnungszeiten und den zusätzlichen Speicherbedarf.
Transit-Node Routing hält den Rekord für die schnellsten Suchzeiten: Diese
sind mehr als eine Million mal schneller als die von Dijkstras Algorithmus. Die Highway Hierarchien verfügen über vergleichsweise niedrige Vorberechnungszeiten von ca. 15 Minuten auf unserem 2,0 GHz AMD Opteron.
Eine Variante von Highway-Node Routing kommt mit lediglich 0,7 Byte
zusätzlichem Speicher pro Knoten aus und ist dabei immer noch mehr als
4 000 mal schneller als Dijkstras Algorithmus. Darüber hinaus handelt es
sich beim Highway-Node Routing um eines der ersten Verfahren, die effizient mit Kantengewichtsänderungen in sehr großen Straßennetzen umgehen können.

Auch sehr große Distanztabellen können schnell berechnet werden,
beispielsweise benötigen wir nicht viel mehr als eine Minute, um eine
20 000 × 20 000 Tabelle zu berechnen; das sind weniger als 0,2 μs pro
Tabelleneintrag. Dijkstras Algorithmus würde mehr als zwei Tage für die
gleiche Berechnung in Anspruch nehmen.

www.ingramcontent.com/pod-product-compliance
Lightning Source LLC
La Vergne TN
LVHW022308060326
832902LV00020B/3335